Universitext

B.I. Dundas
M. Levine
P.A. Østvær
O. Röndigs
V. Voevodsky

Motivic Homotopy Theory

Lectures at a Summer School
in Nordfjordeid, Norway, August 2002

 Springer

Bjørn Ian Dundas
Department of Mathematics
University of Oslo
PO Box 1053, Blindern
0316 Oslo
Norway
E-mail: dundas@math.uib.no

Marc Levine
Northeastern University
Department of Mathematics
360 Huntington Avenue
Boston, MA 02115
USA
E-mail: marc@neu.edu

Paul Arne Østvær
Department of Mathematics
University of Oslo
PO Box 1053, Blindern
0316 Oslo
Norway
E-mail: paularne@math.uio.no

Oliver Röndigs
Fakultät für Mathematik
Universität Bielefeld
Postfach 100 131
33501 Bielefeld
Germany
E-mail: oroendig@math.uni-bielefeld.de

Vladimir Voevodsky
School of Mathematics
Princeton University
Princeton, NJ 08540
USA
E-mail: vladimir@math.ias.edu

Editor:

Bjørn Jahren
Department of Mathematics
University of Oslo
Box 1053 Blindern
0316 Oslo
Norway
E-mail: bjoernj@math.uio.no

Mathematics Subject Classification (2000): 14-xx, 18-xx, 19-xx, 55-xx

Library of Congress Control Number: 2006933719

ISBN-10 3-540-45895-6 Springer Berlin Heidelberg New York
ISBN-13 978-3-540-45895-1 Springer Berlin Heidelberg New York

Springer is a part of Springer Science+Business Media
springer.com
© Springer-Verlag Berlin Heidelberg 2007

Typesetting: by the authors and techbooks using a Springer TEX macro package
Cover design: *design & production* GmbH, Heidelberg

Printed on acid-free paper   SPIN: 11776307   46/techbooks   5 4 3 2 1 0

Preface

This book is based on lectures given at a summer school held in Nordfjordeid on the Norwegian west coast in August 2002. In the little town with the spectacular surroundings where Sophus Lie was born in 1842, the municipality, in collaboration with the mathematics departments at the universities, has established the "Sophus Lie conference center". The purpose is to help organizing conferences and summer schools at a local boarding school during its summer vacation, and the algebraists and algebraic geometers in Norway had already organized such summer schools for a number of years. In 2002 a joint project with the algebraic topologists was proposed, and a natural choice of topic was *Motivic homotopy theory*, which depends heavily on both algebraic topology and algebraic geometry and has had deep impact in both fields.

The organizing committee consisted of Bjørn Jahren and Kristian Ranestad, Oslo, Alexei Rudakov, Trondheim and Stein Arild Strømme, Bergen, and the summer school was partly funded by NorFA — Nordisk Forskerutdanningsakademi. It was primarily intended for Norwegian graduate students, but it attracted students from a number of other countries as well.

These summer schools traditionally go on for one week, with three series of lectures given by internationally known experts. Motivic homotopy theory was an obvious choice for one of the series, and, especially considering the diverse background of the participants, the two remaining series were chosen to cover necessary background material from algebraic topology and model categories, and from algebraic geometry. The background lectures were given by Bjørn I. Dundas and Marc Levine, both of whom have done important work in their respective areas in connection with the main topic of the school. Motivic homotopy theory was taught by one of the founders of the subject and certainly its most prominent figure: Vladimir Voevodsky. We were very happy to have such great and inspiring experts come and share their knowledge and insight with a new generation of students.

After the summer school, Dundas and Levine agreed to write up their lecture series for publication, and Voevodsky agreed to let Oliver Röndigs and Paul Arne Østvær write up his. Röndigs and Østvær have also added an extensive appendix with a more detailed discussion of the homotopy theory and model structures involved. In this volume the contributions of Dundas and Levine are presented first, since they contain the prerequisites for Voevodsky's lectures. They are basically independent and can be read in any order, or just referred to while reading the third part, depending on the background of the reader.

Finally, we would like to thank Springer Verlag for offering to publish this book. We apologize that this has taken longer than expected, but now that the lectures are available, our hope is that many students will find it useful and convenient to find both an introduction to the fascinating subject of motivic homotopy theory and the background material in one place.

Oslo, August 2006   *Bjørn Jahren*

Contents

Background from Algebraic Geometry

Voevodsky's Nordfjordeid Lectures: Motivic Homotopy Theory
Vladimir Voevodsky, Oliver Röndigs, Paul Arne Østvær 147

Prerequisites in Algebraic Topology
the Nordfjordeid Summer School
on Motivic Homotopy Theory

Bjørn Ian Dundas

Department of Mathematics, University of Bergen, Johs. Brunsgt. 12, 5008 Bergen, Norway
dundas@math.ntnu.no

Preface

The intention of this note is **NOT** to write an introductory textbook in algebraic topology!! Many excellent sources exist, let me only point to Hatcher's book [7] which is available online.

On the contrary, these notes (and the resulting lectures at the summer school on motivic homotopy theory) attempt to give a quick overview of the parts of algebraic topology, and in particular homotopy theory, which are needed in order to appreciate that side of motivic homotopy theory.

This means that we have to study spaces and spectra, but in a way which allows for new applications and interpretations. Unfortunately, this point of view is not predominant in most textbooks, and so even students with a first course algebraic topology might be hard put when exposed to this material without some background. In particular, we will use simplicial techniques. Good books on basic simplicial stuff include [6] and [17]. Good books on general model category theory include [21] (the original), [4], [9], and [8].

The first chapter gives a quick presentation of the classical situations where homotopy theory is much used. In the second chapter we make a more thorough study of the key example: simplicial sets. The reason I have chosen to use so much time on this particular example is twofold. Firstly, some of the results were chosen since they were going to be used later in Voevodsky's lectures. Secondly, some of the results were chosen since they are typical models for the kind of arguments that are used over and over again in this theory.

Then a short and inadequate presentation of model category theory appears (this actually was even less complete in the lectures since I was pressed for time at this point). Since spectra are so important to the theory and the set-up uses many of the general ideas of model categories, they close the chapter.

The fourth and last chapter gives one approach to motivic homotopy theory. We give a quick presentation of the category of motivic spaces and their spectra from a functorial point of view. Those not caring overly much for coherent smash-products can stay to the simpler theory, also explained. I stress that this is but one of many possible approaches, and is definitely colored by my own preferences. No proofs are provided, and the reader is referred to [3] for this particular approach, or to [12] for another using symmetric spectra (also discussed in [22]).

Prerequisites. The reader is assumed to be familiar with the basic aspects of point-set topology. For instance Chaps. 2 and 3 in [18] will be (more than) enough. Categorical language is used freely, and some readers will find comfort in having a copy of [16] within easy reach.

Caution. The sketch proofs spread around in these notes are only just that. Although they may seem to be worded like complete proofs, there may be claims put forth which in reality can be hard to establish. The intention has not been to give full proofs, but rather to expose the reader to the idea and methods useful for proving results of this type.

0.1 Notational Quirks

- $\mathbf{N}$: the monoid of natural numbers (contains zero).
- $\mathbf{Z} \subseteq \mathbf{Q} \subseteq \mathbf{R} \subseteq \mathbf{C}$: the rings of integers, rationals, reals and complex numbers.
- $\mathcal{E}ns$: the category of sets.
- $\mathcal{A}b$: the category of abelian groups
- If $\mathcal{C}$ is a category and c and d are objects in $\mathcal{C}$, then $\mathcal{C}(c,d)$ is the set of morphisms in $\mathcal{C}$ from c to d.
- If $\mathcal{C}$ is a category then the *opposite category* $\mathcal{C}^{\mathrm{op}}$ is the category with the same objects, but all arrows reversed.
- If $f\colon a \to c$ and $f\colon b \to c$ are two maps in a category with coproducts, then the natural map $a \coprod b \to c$ is called $f+g$.

I

Basic Properties and Examples

In this chapter we present our basic actors: topological spaces, simplicial sets, simplicial abelian groups, spectra, and chain complexes. We concentrate on the formal structures and the connections between them, and postpone most technicalities.

The category of topological spaces serves as a reference category: it is here the notion of homotopy appears, and many results and constructions are most naturally understood in this context. However, both from a technical point of view and from the point of view of extending the techniques into algebraic geometry, it is better to work in the combinatorial alternatives.

The reason algebraic topologists are free to choose combinatorial models for topological spaces is that algebraic topology is only concerned with those phenomena that can be detected by mapping from certain model spaces. These model spaces are typically discs or simple things you can make by gluing discs together (like spheres), and so it turns out that the spaces that "really matter" are those that **can** be made out of gluing discs together. Simplicial sets are just a wonderful way of doing the bookkeeping for all the gluings in such an approach.

After having presented the few facts we need about topological spaces through the eyes of algebraic topology, we move quickly to simplicial sets and abelian groups.

Traditionally simplicial objects are presented in terms of generators and relations, making things hard to remember. With the assumed categorical proficiency of the audience we stick to the alternative functorial approach where all the bookkeeping is taken care of by ordered sets. In this chapter we essentially just introduce concepts and notation, and will come back to the deeper structure of simplicial sets in the next chapter.

The notion of spectra appears classically through homology theories and give a convenient compromise between spaces and chain complexes capturing much of the important action. It is a practical way of expressing the "linearity" of certain invariants, and will reappear again when we start to talk about motivic homology. We will also touch upon them in Chap. III.

1 Topological Spaces

We start our biased study of algebraic topology by focusing upon those aspects of topological spaces that will become important to us later.

As a motivation for much that is to come, we recall the definition of the singular homology of a topological space.

1.1 Singular Homology

The building blocks of homotopy theory are the "standard simplices". Singular homology gives us an opportunity of recalling how these building blocks are to be assembles.

Definition 1.1.1 Let n be a nonnegative integer. The *standard topological n-simplex*, Δ^n, is the subspace

$$\Delta^n = \left\{ (t_0, \ldots, t_n) \in \mathbf{R}^{n+1} \,\middle|\, \begin{array}{l} \sum_{i=0}^{n} t = 1, \text{ and} \\ t_j \in [0,1] \text{ for all } 0 \leq j \leq n \end{array} \right\}$$

of $\mathbf{R}^{n+1}$.

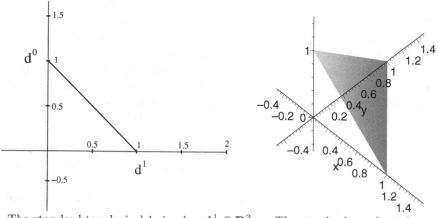

The standard topological 1-simplex $\Delta^1 \subseteq \mathbf{R}^2$. The image of d^0 is the point $(0,1)$, the image of d^1 is $(1,0)$

The standard topological 2-simplex $\Delta^2 \subseteq \mathbf{R}^3$

For each integer i with $0 \le i \le n+1$ the map

$$(t_0, \ldots, t_n) \mapsto (t_0, \ldots, t_{i-1}, 0, t_i, \ldots, t_n)$$

induces an inclusion which we call the ith *face map*

$$d^i \colon \Delta^n \to \Delta^{n+1} .$$

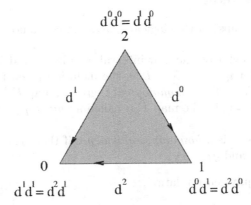

The standard topological 2-simplex Δ^2 seen head on. The images of the d^j's and $d^i d^j$'s are the indicated parts of the boundary. Note that there are relations between the various ways of including 0-simplices (the number attached to each 0-simplex indicates the axis which passes through this point). The *arrows* are only for future reference

For a recollection of the basics on chain complexes, see Sect. 4.1 below. For now, it is enough to recall that a *chain complex* is a sequence

$$(C_*, d) = \left\{ \ldots \xrightarrow{d_{n+1}} C_n \xrightarrow{d_n} C_{n-1} \xrightarrow{d_{n-1}} \ldots \right\}$$

of abelian groups and linear maps such that for all n we have that $d_n d_{n+1} = 0$ (the index runs over the integers). The *homology* of (C_*, d) measures its failure to be exact:

$$H_n(C_*, d) = \frac{\ker\{d_n \colon C_n \to C_{n-1}\}}{\mathrm{im}\{d_{n+1} \colon C_{n+1} \to C_n\}} .$$

Definition 1.1.2 Let X be a topological space. If $n \ge 0$, a *singular n-simplex* is a continuous map $\Delta^n \to X$ from the standard topological n-simplex to X. Let $C_n(X)$ be the free abelian group generated by the singular n-simplices. Define

$$d_n \colon C_n(X) \to C_{n-1}(X)$$

on the generators by letting

$$d_n(\sigma) = \sum_{i=0}^{n} (-1)^i \sigma d^i$$

(we see that $\sigma d^i \colon \Delta^{n-1} \to \Delta^n \to X$ **is** a singular $n-1$-simplex). Note that $dd = 0$. This defines the *singular chain complex* $(C_*(X), d)$, whose homology is called the *singular homology* of X, and is denoted $H_*(X)$.

1.2 Weak Equivalences

From now on, all "maps" of topological spaces are continuous.

Definition 1.2.1 Let I be the unit interval, and let X and Y be topological spaces. For $t \in I$ let $i_t \colon X \to X \times I$ be the inclusion given by $i_t(x) = (x, t)$. Two maps $f_0, f_1 \colon X \to Y$ are *homotopic* if there is a map $F \colon X \times I \to Y$ such that $f_t = Fi_t$ for $t = 0, 1$. The map F is called a *homotopy* from f_0 to f_1 and we write $f_0 \sim f_1$.

A map $f \colon X \to Y$ is a *homotopy equivalence* if there is a map $g \colon Y \to X$ such that $fg \sim 1_Y$ and $gf \sim 1_X$.

Homotopy is an equivalence relation.
We write

$$[X, Y]$$

for the set of homotopy classes of maps from X to Y. If X and Y are pointed (the base point is called $*$ regardless of home) topological spaces we define *pointed homotopy* by insisting that $* \times I \subseteq X \times I$ is sent to $*$. We write

$$[X, Y]_*$$

for the set of **pointed** homotopy classes of maps.

Definition 1.2.2 Let X be a pointed topological space and q a nonnegative integer. Then

$$\pi_q(X) = [S^q, X]_*$$

where the q-sphere $S^q = \{x \in \mathbf{R}^{q+1} \mid |x| = 1\}$ is pointed at $(1, 0, \ldots, 0)$.

The set of path components is exactly $\pi_0(X)$. It is a standard fact that $\pi_q(X)$ is a group (by pinching spheres) for $q \geq 1$ and an abelian group for $q > 1$. They are called the *homotopy groups* of X.

Definition 1.2.3 Let X and Y be topological spaces. A *weak equivalence* is a map $f \colon X \to Y$ such that for any choice of base point f induces an isomorphism

$$\pi_*(X) \to \pi_*(Y)$$

(i.e., it induces an isomorphism $\pi_q(X) \to \pi_q(Y)$ for **all** nonnegative q – and all choices of basepoints).

Fact 1.2.4 If $X \to Y$ is a weak equivalence, then the induced map $H_*(X) \to H_*(Y)$ is an isomorphism too.

A homotopy equivalence is a weak equivalence, but not necessarily conversely. It turns out that for all "reasonable" spaces these notions coincide.

One can show that it makes sense to formally "invert" the weak equivalences. If you do that, you get a category $HoTop$ called the *homotopy category*. We will give an explicit construction later.

1.3 Mapping Spaces

The set of continuous functions $A \to X$ can be given various topologies, but for our purposes the following one is the most convenient one.

Definition 1.3.1 Let A and X be topological spaces. Given a compact subset K of A and an open subset U of X, let

$$W(K, U) = \{ f \mid f \text{ is a continuous function } A \to X \text{ s.t. } f(K) \subseteq U \}$$

The *compact-open* topology on the set of continuous functions from A to X is the topology given by unions of finite intersections of $W(K, U)$'s when K vary over compact subsets of A and U over open subsets of X. The resulting topological space is denoted X^A.

Fact 1.3.2 (see e.g. [18, 46.11]) If A is a locally compact Hausdorff space, and $B, X \in Top$, then there is a natural bijection

$$Top(A \times B, X) \cong Top\left(B, X^A\right) .$$

(Top is the category of topological maps and continuous functions, and so $Top(X, Y)$ is the set of all continuous functions from X to Y).

Conditions like "locally compact" cause havoc in the theory, and is part of the reason why we are searching for combinatorial substitutes for topological spaces. I will try to suppress these issues. For a quick list of relevant facts and references I recommend [9, p. 58ff].

Definition 1.3.3 Let X be a pointed topological space. The *loop space* ΩX on X is the space of **pointed** maps from the standard circle to X.

2 Simplicial Sets

For homotopy theory it suffices to focus on "reasonable" spaces. The so-called CW-complexes are examples of reasonable spaces, but they do not form a nice category. A good substitute, which is often convenient, is simplicial sets. The idea is to build spaces out of standard topological simplices. The only problem is to construct a good category out of the simplices allowing for all the necessary gluings.

Any algebraic geometer would suggest the following: take the "category of standard simplices" and consider the (pre)sheaves on this category. This is exactly what we do, except that we model the simplices by means of finite ordered sets. The connection to topological spaces will be discussed in 2.2 below. Hang on:

2.1 The Category Δ

Definition 2.1.1 Let Δ be the category consisting of the finite ordered sets

$$[n] = \{0 < 1 < 2 < \cdots < n\}$$

for nonnegative integers n, and order-preserving (a.k.a. nondecreasing or weakly monotone) functions.

Definition 2.1.2 A *simplicial set* is a functor

$$\Delta^{\mathrm{op}} \to \mathcal{E}ns$$

where $\mathcal{E}ns$ is the category of sets. A *map of simplicial sets* (or *simplicial map*) is a natural transformation. The category of simplicial sets is denoted $\mathcal{S}$.

More generally, a *simplicial object* in a category $\mathcal{C}$ is a functor $X \colon \Delta^{\mathrm{op}} \to \mathcal{C}$.

If X is a simplicial set we usually we write X_n instead of $X([n])$, and the elements of X_n are referred to as *n-simplices*.

Example 2.1.3 Let n be a nonnegative integer. The *standard (simplicial) n-simplex* $\Delta[n] \in \mathcal{S}$ is given by

$$[q] \mapsto \Delta([q], [n]) = \{\text{order preserving functions } [q] \to [n]\} .$$

Note that the $\Delta[n]$ are nothing but the values of the Yoneda map

$$\Delta \to \mathcal{S}, \qquad [n] \mapsto \Delta[n] = \Delta(-, [n])$$

Note 2.1.4 *By Yoneda's lemma we have a natural isomorphism between X_n and the set of simplicial maps $\Delta[n] \to X$.*

Exercise 2.1.5 Let $k \geq 0$. Show that $\Delta[0]_k = \Delta([k], [0])$ has one element and that $\Delta[1]_k = \Delta([k], [1])$ has $k + 2$ elements. Show that the two maps $[0] \to [1] \in \Delta$ induce injections $\Delta[0] \to \Delta[1]$, and that the union of their images form a simplicial subset $\partial\Delta[1] \subset \Delta[1]$ with two simplices in every dimension. How many k-simplices does the *(simplicial) circle* $S^1 = \Delta[1]/\partial\Delta[1]$ have?

Exercise 2.1.6 Show that there are $\binom{n+1}{k+1}$ **injective** order preserving functions $[k] \to [n]$ (considered as elements in $\Delta[n]_k$ these are called "non-degenerate k-simplices" in $\Delta[n]$.)

A *degenerate* k-simplex in $X \in \mathcal{S}$ is an element $x \in X_k$ such that $x = \phi_* y$ for some $y \in X_n$ and non-injective $\phi \colon [k] \to [n]$.

2.1.7 For the Record: Δ Described by "Generators and Relations"

In particular, for $0 \le i \le n$ we have the maps

$$d^i \colon [n-1] \to [n], \qquad d^i(j) = \begin{cases} j & j < i \\ j+1 & i \le j \end{cases} \qquad \text{"skips } i\text{"}$$

$$s^i \colon [n+1] \to [n], \qquad s^i(j) = \begin{cases} j & j \le i \\ j-1 & i < j \end{cases} \qquad \text{"hits } i \text{ twice".}$$

Every map in Δ has a factorization in terms of these maps. Let $\phi \in \Delta([n], [m])$. Let $\{i_1 < i_2 < \cdots < i_k\} = [m] - im(\phi)$, and $\{j_1 < j_2 < \cdots < j_l\} = \{j \in [n] | \phi(j) = \phi(j+1)\}$. Then

$$\phi(j) = d^{i_k} d^{i_{k-1}} \cdots d^{i_1} s^{j_1} s^{j_2} \cdots s^{j_l}(j) \, .$$

This factorization is unique, and hence we could describe Δ as being generated by the maps d^i and s^i subject to the "cosimplicial identities" :

$$d^j d^i = d^i d^{j-1} \quad \text{for } i < j$$

$$s^j s^i = s^{i-1} s^j \quad \text{for } i > j$$

and

$$s^j d^i = \begin{cases} d^i s^{j-1} & \text{for } i < j \\ id & \text{for } i = j, j+1 \\ d^{i-1} s^j & \text{for } i > j+1 \end{cases} \, .$$

If X is a simplicial set, we let X_n be the image of $[n]$, and for a map $\phi \in \Delta$ we will often write ϕ^* for $X(\phi)$. For the particular maps d^i and s^i, we write simply d_i and s_i for $X(d^i)$ and $X(s^i)$, and call them *face* and *degeneracy maps*. Note that the face and degeneracy maps satisfy the "simplicial identities" which are the duals of the cosimplicial identities.

Hence a simplicial set is often defined in the literature to be a sequence of sets X_n and maps d_i and s_i

satisfying the simplicial identities.

2.2 Simplicial Sets vs. Topological Spaces

We mentioned that Δ was modeled on simplices. This is manifested in the functor $\Delta \to \mathcal{T}op$ given by $[n] \mapsto \Delta^n$ and sending $\phi \colon [n] \to [m] \in \Delta$ to $\phi_* \colon \Delta^n \to \Delta^m$ given by

$$\phi_*(x_0,\ldots,x_n) = \left(\sum_{j \in \phi^{-1}(0)} x_j, \ldots, \sum_{j \in \phi^{-1}(m)} x_j \right)$$

(the face and degeneracies are thus

$$d^i(x_0,\ldots,x_{n-1}) = (x_0,\ldots,x_i,0,x_{i+1},\ldots,x_{n-1})$$
$$s^i(x_0,\ldots,x_{n+1}) = (x_0,\ldots,x_{i-1},x_i + x_{i+1}, x_{i+2}, \ldots, x_{n+1}).$$

Note that the formula "$dd = 0$" for the singular chain complex 1.1.2 follows from the cosimplicial identities).

Exercise 2.2.1 Prove that $[n] \mapsto \Delta^n$ is a functor.

By the way, a functor from Δ to some category is called a *cosimplicial object* in that category.

There is a pair of functors

$$\mathcal{T}op \overset{|-|}{\underset{\text{sing}}{\leftrightarrows}} \mathcal{S}$$

defined as follows.

Definition 2.2.2 For $Y \in \mathcal{T}op$, the *singular complex* is defined as

$$\text{sing}\, Y = \{[n] \mapsto \mathcal{T}op(\Delta^n, Y)\}$$

(the set of unbased continuous functions from the topological standard n-simplex to Y). As $[n] \mapsto \Delta^n$ is a cosimplicial space, this becomes a simplicial set.

Exercise 2.2.3 Elaborate: why is $\text{sing}\, Y$ a simplicial set, and why is sing a functor?

Definition 2.2.4 For $X \in \mathcal{S}$, the *realization* of X is defined as the quotient space

$$|X| = \left(\coprod_n X_n \times \Delta^n \right) / \sim$$

where if $(x,u) \in X_m \times \Delta^n$ and $\phi \colon [n] \to [m] \in \Delta$ we identify the points $(\phi^* x, u) \in X_n \times \Delta^n$ and $(x, \phi_* u) \in X_m \times \Delta^m$.

Example 2.2.5 Several familiar topological spaces are realizations of simplicial sets. Here are some examples:

- The standard topological n-simplex: $\Delta^n \cong |\Delta[n]|$.
- Let $\partial\Delta[n]$ be the sub-simplicial set of $\Delta[n]$ generated by all the *faces* (i.e., the images of the injections $\Delta[n-1] \to \Delta[n]$ induced by the projections $d^i \colon [n] \to [n-1] \in \Delta$, $i = 0, \ldots, n$, cf. 2.1.7). Then $|\partial\Delta[n]|$ is homeomorphic to the $n-1$-sphere.
  In particular, the realization of the (simplicial) circle S^1 of Exercise 2.1.5 is (homeomorphic to) the usual (topological) circle.

Exercise 2.2.6 In the previous example: what does "generate" mean precisely? Prove that $|\partial\Delta[n]|$ is homeomorphic to the $(n-1)$-sphere.

Note 2.2.7 *The picture for $|X|$ is as follows: for each m-simplex $x \in X_m$, you insert a topological m-simplex Δ^m. The maps in Δ keep track of how these simplices should be glued together.*

For instance $(d_i x, u) \sim (x, d^i u)$ tells you that the Δ^{m-1} associated with $d_i x \in X_{m-1}$ should be glued to the ith face of the Δ^m associated with $x \in X_m$.

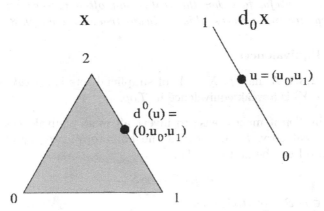

The realization functor glues simplices together. Here $(x, d^0 u) \sim (d_0 x, u)$ tells us that we should identify the two black dots

It is confusing at first to understand the rôle of the surjective maps in Δ. They dictate that once you have a simplex, there will be higher dimensional simplices which are to be identified with it after realization. Often one can do without them (e.g., the singular homology only saw the d_is), but trying to do so in general leads to a much more complex theory. So we just will have to live with facts like that the "interval" $\Delta[1]$ has three 1-simplices: one degenerate for each end point and one spanning the interval itself.

Fact 2.2.8 If X is a simplicial set, then $|X|$ is a compactly generated Hausdorff space, but it is usually not locally compact.

Lemma 2.2.9 *The realization functor is left adjoint to the singular functor.*

Sketch Proof: Check it by hand as an exercise: Explicitly, you have to show that there is a natural bijection of morphism sets

$$\mathcal{T}op(|X|, Y) \cong \mathcal{S}(X, \text{sing}\, Y) \,.$$

This bijection is induced by the "adjunction maps"

$$X \quad \rightarrow \quad \text{sing}\, |X|$$

$$x \in X_n \quad \mapsto \quad (\Delta^n \overset{u \mapsto (x,u)}{\rightarrow} X_n \times \Delta^n \rightarrow |X|) \in \text{sing}\, |X|_n$$

and

$$|\text{sing}\, Y| \quad \rightarrow \quad Y$$

$$(y, u) \in \text{sing}\, (Y)_n \times \Delta^n \quad \mapsto \quad y(u) \in Y. \qquad \qquad \square$$

Note 2.2.10 *That the realization is the left adjoint implies by general nonsense that it preserves all colimits ($\mathcal{S}$ has all (small) (co)limits: just take the (co)limit in each degree separately and you get the right answer).*

It is a very useful fact that the realization often preserves finite limits (when interpreted in "k-spaces" this is always true, see e.g., [9, 3.2.4]).

2.3 Weak Equivalences

Definition 2.3.1 A map $f\colon X \to Y$ of simplicial sets is a *weak equivalence* if $|f|\colon |X| \to |Y|$ is a weak equivalence in $\mathcal{T}op$.

Just as in $\mathcal{T}op$ it makes sense to invert all weak equivalences in $\mathcal{S}$, and the resulting category, $Ho\mathcal{S}$, is called the homotopy category of $\mathcal{S}$. In the simplicial case I can be quite specific:

$$ob\, Ho\mathcal{S} = ob\, \mathcal{S} \,,$$

and if $X, Y \in ob\, \mathcal{S} = ob\, Ho\mathcal{S}$, then

$$Ho\mathcal{S}(X, Y) \cong [|X|, |Y|] \,.$$

This could be a perfectly fine definition. The important point is that the answer is relevant.

Theorem 2.3.2 *(see any book on simplicial sets). The maps of adjunction $X \to \text{sing}\, |X|$ and $|\text{sing}\, Y| \to Y$ are both weak equivalences.*

The singular/realization adjoint pair induces an equivalence between the homotopy categories of topological spaces and simplicial sets.

Note 2.3.3 *We will later see that the correspondence between $\mathcal{S}$ and $\mathcal{T}op$ is even better than just something inducing an equivalence on the homotopy category level (it is something called a Quillen equivalence, see III.1.3.1).*

3 Some Constructions in $\mathcal{S}$

One of the good things about the category of simplicial sets is that it offers no categorical surprises (as opposed to $\mathcal{T}op$ which is just a mess). Limits and colimits are calculated in every degree separately, e.g.,

$$\left(X \coprod Y\right)_q = X_q \coprod Y_q, \text{ and } (X \times Y)_q = X_q \times Y_q .$$

The *mapping space*

$$\underline{\mathcal{S}}(X,Y) = \{[q] \mapsto \mathcal{S}(X \times \Delta[q], Y)\}$$

satisfies the exponential formula

$$\underline{\mathcal{S}}(X, \underline{\mathcal{S}}(Y,Z)) \cong \underline{\mathcal{S}}(Y \times X, Z)$$

on the nose.

Exercise 3.0.1 Prove this.

Note that $\mathcal{S}(X,Y) = \underline{\mathcal{S}}(X,Y)_0$, and that we have a (n associative) "composition"

$$\underline{\mathcal{S}}(Y,Z) \times \underline{\mathcal{S}}(X,Y) \to \underline{\mathcal{S}}(X,Z)$$

induced by sending $g \colon Y \times \Delta[q] \to Z$ and $f \colon X \times \Delta[q] \to Y$ to the composite

$$X \times \Delta[q] \xrightarrow{id \times \text{diagonal}} X \times \Delta[q] \times \Delta[q] \xrightarrow{f \times id} Y \times \Delta[q] \xrightarrow{g} Z$$

3.0.2 Path Components and Homotopies

Let X be a simplicial set, and let $x_0, x_1 \in X_0$ be two zero simplices (a.k.a.. *points* or *vertices*). If there is a 1-simplex $x \in X_1$ such that $d_0 x = x_0$ and $d_1 x = x_1$ we say that x is a *path* from x_0 to x_1. Being connected by a path is neither symmetric nor transitive, but we may anyhow consider the equivalence relation generated by the path relation. We call the set of equivalence classes the set of *path components* and write $\pi_0 X$. We see that $\pi_0 X$ is naturally isomorphic to $\pi_0|X|$.

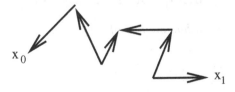

Two points are in the same path components if they can be joined by a finite "chain of paths"

We will later also define the higher homotopy groups in a combinatorial fashion, but whatever definition you use, it should be naturally isomorphic to $\pi_*| - |$.

Definition 3.0.3 Let $X, Y \in \mathcal{S}$, and let f_0 and f_1 be two maps from X to Y. A *homotopy* from f_0 to f_1 is a path in $\underline{\mathcal{S}}(X, Y)$ from f_0 to f_1.

In other words, a homotopy is a map $H\colon X \times \Delta[1] \to Y$ such that the composites

$$X \cong X \times \Delta[0] \xrightarrow{id \times d_i} X \times \Delta[1] \xrightarrow{H} Y, \qquad i = 0, 1$$

are f_1 and f_0.

Since $|X \times \Delta[1]| \cong |X| \times |\Delta[1]|$, we see that the realization of a homotopy is a homotopy in $\mathcal{T}op$.

Definition 3.0.4 A map $f\colon X \to Y \in \mathcal{S}$ is a *(simplicial) homotopy equivalence* if there is a map $g\colon Y \to X$ such that both composites $f \circ g$ and $g \circ f$ are homotopic to the identity.

4 Simplicial Abelian Groups

Let $\mathcal{A}b$ be the category of abelian groups. Since $\mathcal{A}b$ has all (co)limits, so has the category $\mathcal{A} = s\mathcal{A}b$ of simplicial abelian groups. There are free/forgetful functors

$$\mathcal{A}b \overset{\mathbf{Z}[-]}{\underset{U}{\leftrightarrows}} \mathcal{E}ns$$

where $\mathbf{Z}[X]$ is the free abelian group on the set X and UM is the underlying set of the abelian group M. Applying this in every degree we get an adjoint pair

$$\mathcal{A} \overset{\mathbf{Z}[-]}{\underset{U}{\leftrightarrows}} \mathcal{S} .$$

Note 4.0.5 *The category $\mathcal{A} = s\mathcal{A}b$ inherits structure from $\mathcal{S}$ via this adjoint pair. For instance $\mathcal{A}$ has its own "internal" hom-object via*

$$\underline{\mathcal{A}}(M, N)_q = \mathcal{A}(M \otimes \mathbf{Z}[\Delta[q]], N) ,$$

and so a natural notion of homotopies of maps.

4.1 Simplicial Abelian Groups vs. Chain Complexes

As usual, a chain complex is a sequence

$$C_* = \{\cdots \leftarrow C_{q-1} \leftarrow C_q \leftarrow C_{q+1} \leftarrow \ldots\}$$

such that any composite is zero, and a map of chain complexes $f_* \colon C_* \to D_*$ is a collection of maps $f_q \colon C_q \to D_q$ such that the diagrams

$$
\begin{array}{ccc}
C_q & \xrightarrow{\;f_q\;} & D_q \\
\downarrow & & \downarrow \\
C_{q-1} & \xrightarrow{\;f_{q-1}\;} & D_{q-1}
\end{array}
$$

commute. We let Ch be the category of chain complexes, and $Ch^{\geq 0}$ be the full subcategory of chain complexes C_* such that $C_q = 0$ if $q < 0$.

If C_* is a chain complex, we let

$$
\begin{aligned}
Z_q C &= \ker\{C_q \to C_{q-1}\} \ (\text{cycles}), \\
B_q C &= \operatorname{im}\{C_{q+1} \to C_q\} \ (\text{boundaries}) \text{ and} \\
H_q C_* &= Z_q C / B_q C \ (\text{homology}).
\end{aligned}
$$

A simplicial abelian group M gives rise to a chain complex called the *Moore complex*:

$$C_*(M) = \{M_0 \xleftarrow{\;d_0 - d_1\;} M_1 \xleftarrow{\;d_0 - d_1 + d_2\;} M_2 \xleftarrow{\;d_0 - d_1 + d_2 - d_3\;} \ldots\}.$$

Note that the singular homology of a topological space Y 1.1.2 was defined as

$$H_*(Y) = H_*(C_* \mathbf{Z}[\operatorname{sing} Y]).$$

More generally, for a simplicial set X we define

$$H_*(X) = H_*(C_* \mathbf{Z}[X]).$$

Fact 4.1.1 Let M be a simplicial abelian group. Then there is a natural isomorphism

$$\pi_*(|U(M)|) \cong H_*(C_*(M)).$$

The map $X \cong 1 \cdot X \subseteq U\mathbf{Z}[X]$ induces what is called the *Hurewicz map* $\pi_*(|X|) \to H_*(|X|)$ on homotopy groups [7, 4.2].

4.2 The Normalized Chain Complex

If M is a simplicial abelian group, then the normalized chain complex $C_*^{\mathrm{norm}}(M)$ (which is usually called N_*M, an option unpalatable to us since this notation will be occupied by the nerve) is the chain complex given by

$$C_q^{\mathrm{norm}}(M) = \bigcap_{i=0}^{q-1} \ker\{d_i : M_q \to M_{q-1}\}$$

and boundary map $C_q^{\mathrm{norm}} M \to C_{q-1}^{\mathrm{norm}} M$ given by the remaining face map d_q.

Fact 4.2.1 (Dold-Kan) The normalized chain complex gives an equivalence of categories

$$C^{\mathrm{norm}} : \mathcal{A} \to Ch^{\geq 0} .$$

Furthermore, this equivalence sends homotopies to chain homotopies and vice versa for the adjoint.

The inclusion of the normalized complex into the Moore complex $C_*^{\mathrm{norm}}(M) \subseteq C_*(M)$ is a homotopy equivalence (see e.g., [6, III.2]).

5 The Pointed Case

Most of what have been said so far carries over to the pointed setting (a pointed simplicial set is by definition a simplicial pointed set in case you wondered). Just a tad of notational stuff. We write $\mathcal{S}_*$ for the category of pointed simplicial sets.

If X and Y are pointed sets, then the *wedge*

$$X \vee Y$$

is what you get if you take the disjoint union of X and Y and identify their basepoints. Alternatively (and more concretely) you may think of $X \vee Y$ as the subset of $X \times Y$ where (at least) one of the coordinates is the base point. The quotient

$$X \wedge Y = (X \times Y)/(X \vee Y)$$

is called the *smash*. The *suspension* of $X \in \mathcal{S}_*$ is just another word for the smash $S^1 \wedge X$ where $S^1 = \Delta[1]/\partial\Delta[1]$ is the simplicial circle. We define the higher spheres by

$$S^n = S^1 \wedge S^{n-1} ,$$

giving the isomorphisms $S^n \wedge S^m \cong S^{m+n}$. By the way, $S^0 = \{0,1\} = \partial\Delta[1]$ pointed in 0.

If Z is an (unpointed) set, then

$$Z_+$$

is Z to which we have added a base point.

The free/forgetful pair connecting abelian groups and sets factors through the pointed sets (abelian groups are pointed in zero)

$$Ab \underset{\tilde{\mathbf{z}}[-]}{\leftrightarrows} \mathcal{E}ns_* \underset{X \mapsto X_+}{\leftrightarrows} \mathcal{E}ns$$

where $\tilde{\mathbf{Z}}[X] = \mathbf{Z}[X]/\mathbf{Z}[*]$, $\mathcal{E}ns_*$ is the category of pointed sets, and the arrows pointing to the right are the forgetful functors (on the bottom, following the convention that right adjoints are written below the left adjoint).

There are many (natural!) relations between these constructions, such as

1. $(Z \times S)_+ \cong Z_+ \wedge S_+$,
2. $X \wedge (Y_1 \vee Y_2) \cong (X \wedge Y_1) \vee (X \wedge Y_2)$,
3. $S^0 \wedge X \cong X$,
4. $\tilde{\mathbf{Z}}[X \vee Y] \cong \tilde{\mathbf{Z}}[X] \oplus \tilde{\mathbf{Z}}[Y]$,
5. $\tilde{\mathbf{Z}}[X \wedge Y] \cong \tilde{\mathbf{Z}}[X] \otimes \tilde{\mathbf{Z}}[Y]$
6. if $A \subseteq X$, then $0 \to \tilde{\mathbf{Z}}[A] \to \tilde{\mathbf{Z}}[X] \to \tilde{\mathbf{Z}}[X/A] \to 0$ is exact.

Exercise 5.0.1 Write up a couple more relations like the ones above and prove all of them. Note that the short exact sequence in 6 splits (if you believe in the axiom of choice); but this splitting will **not** be natural (regardless of faith).

Performing these constructions degreewise, we get the corresponding constructions for pointed simplicial sets.

Reduced homology is given as the homotopy groups of $\tilde{\mathbf{Z}}[X]$, the wedge axiom is reflected in 4, the Künneth theorem in 5, and excision in 6.

Definition 5.0.2 The space $\tilde{\mathbf{Z}}[S^n]$ is called the *nth integral Eilenberg-Mac Lane space*, and has the property that

$$\pi_q \tilde{\mathbf{Z}}[S^n] = \begin{cases} 0 & \text{if } n \neq q \\ \mathbf{Z} & \text{if } n = q \end{cases}.$$

5.0.3 Mapping Spaces and Homotopies

We have mapping spaces as well, let X and Y be pointed simplicial sets, then $\underline{S}_*(X, Y)$ is the pointed simplicial set with q simplices

$$\underline{S}_*(X, Y)_q = \{\text{pointed simplicial maps } X \wedge \Delta[q]_+ \to Y\}.$$

Exercise 5.0.4 Note that $S_*(X, Y) = \underline{S}_*(X, Y)_0$, and that we have a "composition"

$$\underline{S}_*(Y, Z) \wedge \underline{S}_*(X, Y) \to \underline{S}_*(X, Z),$$

and an exponential law

$$\underline{S}_*(X \wedge Y, Z) \cong \underline{S}_*(Y, \underline{S}_*(X, Z))$$

If $f_0, f_1 \in S_*(X, Y)$, a *(pointed) homotopy* between them is a path in $\underline{S}_*(X, Y)$. In other words, it is a map

$$X \wedge \Delta[1]_+ \to Y$$

restricting to f_0 and f_1 on the boundary of $\Delta[1]$. (Pointed) *homotopy equivalences* are defined in the obvious way.

5.0.5 Loop Spaces and Cohomology

We note that $\operatorname{sing} \Omega|X| \cong \underline{\mathcal{S}}_*(S^1, \operatorname{sing}|X|)$ (this uses 1.3.2, that the circle is (locally) compact Hausdorff and that realization commutes with finite products if one of the factors is locally compact).

It is an important fact that even though $X \to \operatorname{sing}|X|$ is a weak equivalence,

$$\underline{\mathcal{S}}_*(A, X) \to \underline{\mathcal{S}}_*(A, \operatorname{sing}|X|)$$

may not be a weak equivalence.

Exercise 5.0.6 Prove that $\pi_0 \underline{\mathcal{S}}_*(S^1, S^1) \cong S^0$, whereas $\pi_0 \underline{\mathcal{S}}_*(S^1, \operatorname{sing}|S^1|) \cong \pi_1|S^1| \cong \mathbf{Z}$.

It is $\underline{\mathcal{S}}_*(A, \operatorname{sing}|X|)$ which has the "right" homotopy type, and we define the *loop space*

$$\Omega X = \underline{\mathcal{S}}_*(S^1, \operatorname{sing}|X|) \cong \operatorname{sing} \Omega|X| \ .$$

Then loop and suspension are not adjoint, but we still get that a map $S^1 \wedge X \to Y$ induces a map $X \to \Omega Y$ (by using the adjunction on the composite $S^1 \wedge X \to Y \to \operatorname{sing}|Y|$).

Definition 5.0.7 Let $X \in \mathcal{S}$. The nth (reduced, integral) *cohomology* of X is given by the group

$$\tilde{H}^n(X) = \pi_0 \underline{\mathcal{S}}_*(X, \tilde{\mathbf{Z}}[S^n])$$

Note 5.0.8 *It is a special feature of $\tilde{\mathbf{Z}}[S^n]$ that there is no need for $\operatorname{sing}|-|$ around it (since it is a simplicial abelian group it is "fibrant" in a language to come).*

Once this point is properly understood, it is not too difficult to derive the axioms for cohomology from this definition. A direct proof that it agrees with the usual cochain definition (not given here) is an application of the isomorphism of the category of simplicial abelian groups and chain complexes concentrated in non-negative degrees (together with the fact that $\tilde{\mathbf{Z}}[S^n]$ corresponds to the chain complex with a single nontrivial group $\mathbf{Z}$ concentrated in degree n).

6 Spectra

6.1 Introduction

Many phenomena and invariants are "stable" in the sense that suspending simply acts as shifting. More to the point: for (generalized) cohomology theories, gluing of spaces can be easily understood through excision, and Brown representability says that any cohomology theory can be realized by mapping into a "spectrum". Spectra are a sensible half-way house between spaces and abelian groups where cohomology theories "live"; here suspension is equivalent to shifting and finite coproducts are equivalent to products. However, the category of spectra is not simply a jumble of invariants with values in abelian groups, it carries the same kind of structure as the category of spaces and is open for analysis through the same type of machinery.

Definition 6.1.1 In algebraic topology (as opposed to algebraic geometry), a *spectrum* is a sequence of simplicial sets

$$E = \{E^0, E^1, E^2, \dots\}$$

together with (structure) maps

$$S^1 \wedge E^k \to E^{k+1}$$

for $k \geq 0$. A map of spectra $f \colon E \to F$ is a sequence of maps $f^k \colon E^k \to F^k$ compatible with the structure maps: the diagrams

$$
\begin{array}{ccc}
S^1 \wedge E^k & \longrightarrow & E^{k+1} \\
\downarrow {\scriptstyle id_{S^1} \wedge f^k} & & \downarrow {\scriptstyle f^{k+1}} \\
S^1 \wedge F^k & \longrightarrow & F^{k+1}
\end{array}
$$

commute. We let $\mathcal{S}pt$ be the resulting category of spectra.

This definition is apparently due to Lima [14].
     There are some especially important spectra:

Example 6.1.2   1. the *sphere spectrum*

$$\underline{\mathbf{S}} = \{k \mapsto S^k = S^1 \wedge \cdots \wedge S^1\}$$

whose structure maps are the identity.
   2. the (integral) *Eilenberg-Mac Lane spectrum*

$$H\mathbf{Z} = \{k \mapsto \tilde{\mathbf{Z}}[S^k]\}$$

whose structure map is induced by the natural map $\tilde{\mathbf{Z}}[X] \wedge Y \to \tilde{\mathbf{Z}}[X \wedge Y]$.

The Eilenberg-Mac Lane spectra are examples of Ω-*spectra*, that is the adjoint of the structure maps give rise to equivalences $E^k \to \Omega E^{k+1}$. In various treatments this property is taken as a part of the definition of spectra. We do not; it takes more categorical effort to make this work. Our approach is a typical example of how modern homotopy theory treat this kind of issues. The Ω-spectra are admittedly the spectra that matter, but many natural constructions on spectra takes us outside the Ω-spectra, and so the approach is to admit all spectra, but allow for equivalences (*stable equivalences*, see below) that are measured by Ω-spectra (in model categorical language to be explained in Chap. III, the Ω-spectra are the *fibrant* spectra).

6.2 Relation to Simplicial Sets

If X is a pointed simplicial set and E is a spectrum, then $E \wedge X$ is the spectrum $n \mapsto E^n \wedge X$, and E^X is the spectrum $n \mapsto \underline{S}_*(X, E^n)$.

There is a pair of adjoint functors

$$Spt \underset{R}{\overset{\Sigma^\infty}{\rightleftarrows}} S_*$$

where

$$\Sigma^\infty X = \{n \mapsto S^n \wedge X\}$$

is the *suspension spectrum* with right adjoint $RE = E^0$ – the *zeroth space*.

Occasionally RE is referred to as the "underlying space" of the spectrum, but this term is also sometimes used for the *underlying infinite loop space* $\Omega^\infty E = \lim_{\overrightarrow{n}} \Omega^n E^n$ (the maps in the colimit are the same as in 6.3.1 below).

Exercise 6.2.1 Prove that R and Σ^∞ are adjoint.

6.3 Stable Equivalences

The relevant equivalences giving the right correspondence between cohomology theories and spectra are the stable equivalences:

Definition 6.3.1 Let E be a spectrum. The (stable) *homotopy groups* of E are defined as

$$\pi_q E = \lim_{\overrightarrow{k}} \pi_{q+k} E^k$$

where the colimit is over the maps $\pi_{q+k} E^k \to \pi_{q+k} \Omega E^{k+1} \cong \pi_{q+k+1} E^{k+1}$ for $k > -q$.

Exercise 6.3.2 Make the maps $\pi_{q+k} E^k \to \pi_{q+k} \Omega E^{k+1} \cong \pi_{q+k+1} E^{k+1}$ explicit using the definition in 5.0.5 of Ω in S_*.

Note that the stable homotopy groups define a functor from spectra to **Z**-graded abelian groups.

Definition 6.3.3 A map of spectra $f \colon E \to F$ is a *stable equivalence* if it induces an isomorphism on stable homotopy groups.

Fact 6.3.4 An important fact about spectra is that the natural map

$$E \vee F \to E \times F$$

is a stable equivalence. This is related to the fact that if $X, Y \in S$ are such that $\pi_i X = 0$ for $i < n$ and $\pi_j Y = 0$ for $j < m$, then $\pi_k(X \wedge Y) = 0$ for $k < m + n$.

Exercise 6.3.5 In the "fact" above: how do you define $E \vee F$ and $E \times F$ (i.e., what are the structure maps)?

Again it makes sense to invert all stable equivalences, and the resulting category is called the *stable homotopy category*,

$$\mathcal{H}o\mathcal{S}pt$$

(often in the literature you will find the term "the stable category", a term which may cause confusion, and which we will avoid). Facts like 6.3.4 adds up to show that $\mathcal{H}o\mathcal{S}pt$ is an additive category (finite sum = product), and even better, it has a "tensor product" which is somehow derived from the smash on $\mathcal{S}_*$.

6.4 Homology Theories

A spectrum E gives rise to a "*(co)homology theory*": if X is a simplicial set, we let

$$E_n(X) = \pi_n(E \wedge X)$$

and

$$E^n(X) = \pi_{-n} E^X .$$

The *stable homotopy group* $\pi_n^S(X)$ of a pointed space X is by definition

$$\underline{S}_n(X) = \pi_n(\underline{S} \wedge X) = \varinjlim_k \pi_{n+k}(S^k \wedge X) .$$

Theorem 6.4.1 *There are natural isomorphisms*

$$(H\mathbf{Z})_n(X) \cong \tilde{H}_n(X)$$

$$(H\mathbf{Z})^n(X) \cong \tilde{H}^n(X) .$$

Sketch Proof: Homology part: First note that $\tilde{H}_n(X) \cong \tilde{H}_{n+k}(S^k \wedge X) \cong \pi_{n+k} \tilde{\mathbf{Z}}[S^k \wedge X]$ for all $k \geq 0$. Given this, the natural isomorphism $(H\mathbf{Z})_n(X) \cong \tilde{H}_n(X)$ is given as the colimit (over k) of

$$\pi_{n+k}((H\mathbf{Z} \wedge X)^k) = \pi_{n+k}(\tilde{\mathbf{Z}}[S^k] \wedge X) \to \pi_{n+k}(\tilde{\mathbf{Z}}[S^k \wedge X])$$

which is an isomorphism for $k > n$ (by a "stability result" similar to 6.3.4).

Cohomology part: $\tilde{H}^n(X) = \tilde{H}^k(S^{k-n} \wedge X) = \pi_0 \underline{S}_*(S^{k-n} \wedge X, \tilde{\mathbf{Z}}[S^k]) \cong \pi_{k-n} \underline{S}_*(X, \tilde{\mathbf{Z}}[S^k])$ for all $k \geq n$. □

6.5 Relation to Chain Complexes

There is a close connection between chain complexes and spectra.

The definition of spectra is reminiscent of how you'd represent arbitrary chain complexes by means of chain complexes concentrated in non-negative dimensions: a chain complex C can be given by a sequence

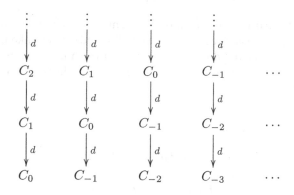

of non-negatively graded chain complexes C^0, C^1,... together with isomorphisms $C^i_j \cong C^{i+1}_{j+1}$. Homology in arbitrary dimensions is then accessible as

$$H_j(C) \cong \varinjlim_n H_{n+j}(C^n).$$

Notice that if we let $\mathbf{Z}[1]$ be the chain complex concentrated in degree 1 with a single $\mathbf{Z}$, then the isomorphism $C^i_j \cong C^{i+1}_{j+1}$ can be reformulated through a map

$$\mathbf{Z}[1] \otimes C^i \to C^{i+1}$$

which is an isomorphism in positive degrees. The tensor product of chain complexes is given by $(C \otimes D)_n = \oplus_{p+q=n} C_p \otimes D_q$ (with the appropriate sign conventions on the differentials), and so $(\mathbf{Z}[1] \otimes C^i)_n \cong C^i_{n-1} \cong C^{i+1}_n$ for $n > 0$.

As a matter of fact, if we replace the category of simplicial sets $\mathcal{S}$, the smash product $\wedge$ and the circle S^1 with

1. the category of simplicial abelian groups $\mathcal{A}$, with degreewise tensor, and $\tilde{\mathbf{Z}}[S^1]$
2. the category of chain complexes concentrated in non-negative degrees $Ch^{\geq 0}$, with tensor of chain complexes and $C^{\text{norm}}\tilde{\mathbf{Z}}[S^1] = \mathbf{Z}[1]$ or
3. the category of chain complexes Ch, with tensor of chain complexes and $\mathbf{Z}[1]$

word for word in the definition of spectra we get categories $\mathcal{S}pt(\mathcal{A})$, $\mathcal{S}pt(Ch^{\geq 0})$ and $\mathcal{S}pt(Ch)$ which for all practical purposes play the rôle of chain complexes, and which is related to spectra by the usual free/forgetful functor connecting abelian groups and sets.

The relation $\mathcal{S}pt$ to Ch is as follows

$$\mathcal{S}pt \overset{\tilde{\mathbf{Z}}[-]}{\underset{}{\leftrightarrows}} \mathcal{S}pt(\mathcal{A}) \underset{C^{\mathrm{norm}}}{\leftrightarrows} \mathcal{S}pt(Ch^{\geq 0}) \overset{\mathrm{truncate}}{\underset{\mathrm{include}}{\leftrightarrows}} \mathcal{S}pt(Ch) \underset{R}{\leftrightarrows} Ch$$

The maps to the left of $\mathcal{S}pt(\mathcal{A})$ all induce equivalences on the associated homotopy categories.

II

Deeper Structure: Simplicial Sets

In this chapter we will develop some further properties necessary to understand simplicial sets. In order to control the weak equivalences we introduce two classes of maps: fibrations and cofibrations. These maps formalize "obstruction theory", or rather they tell us when existence of liftings can be expected. This is intimately connected with the fact that weak equivalences are not isomorphisms, although they become so in the homotopy category.

So a natural question could be: for what kind of weak equivalences $X \xrightarrow{\sim} Y$ can we expect to find "liftings" for each diagram

$$
\begin{array}{ccc}
A & \longrightarrow & X \\
\text{\scriptsize injective}\downarrow & & \downarrow\simeq \\
B & \longrightarrow & Y
\end{array}
$$

i.e., maps $B \to X$ you can insert in the diagram without destroying the commutativity? This is reminiscent of Tietze's extension theorem in point set topology which states that if $X = [0,1] \xrightarrow{\sim} * = Y$ then extensions exist for all closed inclusions $A \subseteq B$ where B is normal. What makes topological spaces different from simplicial sets is that "cofibrations" (more about these later) in $\mathcal{T}op$ are rather complicated, making qualifications such as "closed" and "normal" necessary. By contrast a "cofibration" of simplicial sets is simply an inclusion.

By choosing $A \subseteq B$ to be $\emptyset \subseteq Y$ we see that such an equivalence $X \xrightarrow{\sim} Y$ would have a splitting $Y \to X$, and by another choice of $A \to B$ (*hint*: try to lift a "trivial homotopy") we can show that $X \xrightarrow{\sim} Y \to X$ is homotopic to the identity. So in particular $X \to Y$ is a homotopy equivalence, and $Y \to X$ realizes its homotopy inverse.

A systematizing fact about simplicial sets is that they can be built by gluing simplices along their boundary, so the inclusions $\partial\Delta[n] \subseteq \Delta[n]$ play a prominent rôle. Here $\partial\Delta[n]$ is the subcomplex of $\Delta[n]$ generated by the faces d^i.

The inclusions that are also weak equivalences also have their "building blocks" discussed later, and the fibrations alluded to above are the maps $X \to Y$ having the property that for every diagram

$$
\begin{array}{ccc}
A & \longrightarrow & X \\
\downarrow & & \downarrow \\
B & \longrightarrow & Y
\end{array}
$$

where $A \to B$ is both an inclusion and a weak equivalence we have a lifting $B \to X$. The simplicial sets X having the property that the canonical map $X \to *$ is a fibration are called *fibrant*, and play a preferred rôle. Homotopy theory for fibrant objects is not hampered by problems such as weak equivalences not having homotopy inverses.

Any simplicial set is equivalent to a fibrant simplicial set, so why not just stay with the fibrant ones? The reason is that whereas $\mathcal{S}$ has good and transparent categorical properties, the subcategory of fibrant objects is very bad. Many of the constructions we use will take you out of fibrant objects. So the solution is to stick to $\mathcal{S}$, but have the (co)fibrations and weak equivalences as part of your data.

It all ends up in the statement that $\mathcal{S}$ is a "model category" (see Chap. III) and the realization functor is a "Quillen equivalence" $\mathcal{T}op \to \mathcal{S}$. From this technical point of view (which we will develop in the next chapter) the thing which is special about $\mathcal{S}$ (apart from its categorical simplicity) is that all objects are "cofibrant", and properties relating to the fact that π_* commutes with colimits over $\mathbf{N}$.

Before we start with the technical issues we allow ourselves to give a variation of an important point from the first chapter.

0.1 Realization as an Extension Through Presheaves

Another way of presenting the realization/singular adjoint pair, more reminiscent of the idea of simplicial sets as presheaves is the following. Consider the topological standard simplices as a functor $\Delta \to \mathcal{T}op$, and extend it to simplicial sets through the Yoneda functor

$$
h \colon \Delta \to \text{Presheaves on } \Delta = \mathcal{S}, \qquad [n] \mapsto \Delta[n] = \Delta(-, [n]) \, .
$$

The realization is the "filler" (the precise notion is "left Kan extension" according to most working mathematicians) in

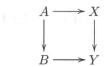

making the diagram commutative up to natural isomorphism (the above mentioned working mathematicians would also write things like $|X| = \int^{[n]} X_n \times \Delta^n$ and claim that the string of symbols

$$\mathcal{T}op\left(\int^{[n]} X_n \times \Delta^n, Y\right) \cong \int_{[n]} \mathcal{T}op(X_n, Y^{\Delta^n}) \cong \int_{[n]} \mathcal{E}ns(X_n, \mathcal{T}op(\Delta^n, Y))$$

is another way of saying that the realization and singular functors are adjoint).

0.1.1 Categories and Simplicial Sets

The ideas involved in the singular/realization pair connecting topological spaces and simplicial sets is quite general, and we will see other examples later. For now we are content with giving just one, namely the connection between small categories and simplicial sets.

The category Δ may be considered as a subcategory of the category $\mathcal{C}at$ of small categories, by viewing $[q] \in \Delta$ as the category $\{0 \leftarrow 1 \leftarrow \cdots \leftarrow q\}$. Order preserving functions correspond to functors, and so we have an "inclusion" $\Delta \hookrightarrow \mathcal{C}at$.

In analogy with the singular functor, the Yoneda functor $h \colon \Delta \to \mathcal{S}$ extends to the *nerve*

$$N \colon \mathcal{C}at \to \mathcal{S} .$$

For a given category $\mathcal{C}$ the q simplices $N_q\mathcal{C}$ of the nerve is the set of functors $[q] \to \mathcal{C}$, or in other words, the set of all composable arrows

$$c_0 \leftarrow c_1 \leftarrow \cdots \leftarrow c_q$$

in $\mathcal{C}$. The nerve embeds $\mathcal{C}at$ as a full subcategory of $\mathcal{S}$.

The left adjoint to the nerve – corresponding to the realization – is less important, but is constructed as before as the filler in

$$
\begin{array}{ccc}
\Delta & \hookrightarrow & \mathcal{C}at \\
h \downarrow & \nearrow & \\
\mathcal{S} &
\end{array}
$$

Exercise 0.1.2 (nerves, natural transformations and homotopies). Let

$$f_0, f_1 \colon \mathcal{C} \to \mathcal{D}$$

be functors between small categories. Prove that there is a natural transformation from f_0 to f_1 if and only if there is a homotopy from Nf_0 to Nf_1. (*Hint:* $N[1] = \Delta[1]$.)

1 (Co)fibrations

1.1 Simplicial Sets are Built Out of Simplices

Let us try to make some content to the title of this subsection.

I could mean the standard fact that "presheaves are colimits of representables": If X is a simplicial set, let the *simplex category* ΔX be the category of representable objects over X (i.e., the objects of ΔX are maps $\Delta[n] \to X$, and a morphism is a commutative diagram

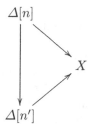

where – by Yoneda – the vertical map is induced by a map $[n] \to [n'] \in \Delta$). We have a functor $\Delta X \to \mathcal{S}$ sending $\Delta[n] \to X$ to $\Delta[n]$, and a natural isomorphism from the colimit of this functor to X.

However, what I really have in mind when writing the title is the following fact:

Lemma 1.1.1 *Let $K \subseteq L$ be an inclusion of simplicial sets. Then there is a functorial factorization*

$$K = K(-1) \subseteq K(0) \subseteq K(1) \subseteq \cdots \subseteq \bigcup_{n \geq -1} K(n) = L$$

such that $K(i-1) \subseteq K(i)$ is gotten by "attaching cells", i.e., it fits in a pushout diagram

$$
\begin{array}{ccc}
\coprod \partial\Delta[i] & \xrightarrow{\;\subseteq\;} & \coprod \partial\Delta[i] \\
\downarrow & & \downarrow \\
K(i-1) & \xrightarrow{\;\subseteq\;} & K(i)
\end{array}
$$

Sketch Proof: By induction, we define $K(i)$ such that $K_n \subseteq L_n$ factors as $K_n \subseteq K(i)_n = L_n$ for all $n \leq i$ as follows. Assume we have constructed $K(i-1)$, and consider the complement $L_i \setminus K(i-1)_i$. By the Yoneda lemma we may consider each element in $L_i \setminus K(i-1)_i$ as a map $\Delta[i] \to L$, and by the assumption on $K(i-1)$ the composite $\partial\Delta[i] \subseteq \Delta[i] \to L$ factors through $K(i-1) \subseteq L$.

Now define $K(i)$ by means of the pushout in the statement of the lemma where the coproduct is taken over $L_i \setminus K(i-1)_i$. Finally one must check that the canonical map $K(i) \to L$ is an injection (and so can be chosen to be an inclusion). $\qquad\square$

1.2 Lifting Properties and Factorizations

In this section we will meet our first argument involving lifting properties. It is important because we look at an example which displays some standard methods, and forms the cornerstone of homotopical algebra.

In homotopy theory one of the important issues are when we can lift or extend maps. For instance, if a map $f\colon \partial\Delta[n] \to X$ in $\mathcal{S}$ can be extended to a map $\Delta[n] \to X$, then $|f|$ does not contribute to the homotopy of $|X|$. More generally, given a (commutative) diagrams of the form

$$
\begin{array}{ccc}
A & \longrightarrow & X \\
{\scriptstyle j}\downarrow & & \downarrow{\scriptstyle f} \\
B & \longrightarrow & Y
\end{array}
$$

one may ask whether there is a map $s\colon B \to X$ making the following diagram commutative

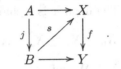

We call such a map s a *lifting* of the original diagram. For instance, the class of maps $f\colon X \to Y$ having the property that all diagrams of the form

$$
\begin{array}{ccc}
\partial\Delta[n] & \longrightarrow & X \\
{\scriptstyle \text{incl.}}\downarrow & & \downarrow{\scriptstyle f} \\
\Delta & \longrightarrow & Y
\end{array}
$$

for $n \geq 0$ have liftings, are called *trivial fibrations* (not to be confused with product fibrations: I am sorry about the unfortunate terminology). We refer to the defining property of trivial fibrations as "the maps having the *right lifting property* with respect to the inclusions $\partial\Delta[n] \subseteq \Delta[n]$", and we will often display trivial fibrations by decorated arrows: $\overset{\sim}{\longrightarrow\!\!\!\!\!\longrightarrow}$.

Note 1.2.1 *In view of Lemma 1.1.1, trivial fibration may be alternatively classified as the maps having the right lifting property with respect to arbitrary injections of simplicial sets.*

Theorem 1.2.2 *Any map $f\colon X \to Y \in \mathcal{S}$ may be factored as an inclusion followed by a trivial fibration*

$$
X \overset{\hookrightarrow}{\longrightarrow} Z \overset{\sim}{\longrightarrow\!\!\!\!\!\longrightarrow} Y \ .
$$

Furthermore, this factorization may be chosen functorially.

Sketch Proof: Consider the set D_1 of diagrams of the form

$$\begin{array}{ccc} \partial\Delta[n] & \longrightarrow & X \\ \text{incl.}\downarrow & & \downarrow f \\ \Delta[n] & \longrightarrow & Y \end{array}$$

Let X_1 be what you get if you "fill all these holes", i.e., the pushout

$$\begin{array}{ccc} \coprod_{d\in D_1}\partial\Delta[n_d] & \longrightarrow & X \\ \coprod_{d\in D_1}\text{incl.}\downarrow & & \downarrow \\ \coprod_{d\in D_1}\Delta[n_d] & \longrightarrow & X_1 \end{array}$$

Note that $X \to X_1$ is an inclusion. By the universal property of the pushout, f factors through $X \to X_1$, and we can play the game over again, this time to the induced map $X_1 \to Y$. The upshot is a chain

$$X = X_0 \subseteq X_1 \subseteq X_2 \subseteq \dots$$

whose colimit (union) we call X_∞. We let

$$X \xrightarrow{\iota_f} Z_f \xrightarrow[\sim]{\phi_f} Y$$

be

$$X \to X_\infty \to Y$$

Note that $X \to X_\infty$ is an inclusion. The important thing is that $X_\infty \to Y$ is a trivial fibration.

To see this, we need to observe that $\partial\Delta[n]$ is *small* (much more about small objects later), which for our current purposes implies that a map $\partial\Delta[n] \to X_\infty$ must actually factor through $X_m \subseteq X_\infty$ for some (possibly very big) integer m.

So if we have a square of the sort

$$\begin{array}{ccc} \partial\Delta[n] & \longrightarrow & X_\infty \\ \text{incl.}\downarrow & & \downarrow \phi_f \\ \Delta[n] & \longrightarrow & Y \end{array}$$

and ask for a lifting, let m be such that $\partial\Delta[n] \to X_\infty$ factors through $X_m \to X_\infty$ and notice that by the very construction of X_{m+1} we have a commutative diagram

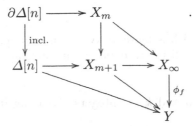

But since $X_{m+1} \to Y$ factors through X_∞ we have our desired lifting.   □

1.3 Small Objects

The proof above exemplified what is known as the *small object argument*.
    The simplest case is well known for sets.

Fact 1.3.1 (see e.g., [16]: "filtered colimits commute with finite limits") If $X_0 \to X_1 \to \dots$ is a sequence of sets, and A a finite set, then the natural function

$$\varinjlim \left(X_i^A \right) \to \left(\varinjlim X_i \right)^A$$

is a bijection.

The full-fledged version used in homotopy theory these days often involves a lot of big cardinals, but this is more than enough for our present purposes.
    A *finite* simplicial set is a simplicial set with only finitely many non-degenerate simplices. Examples include $\Delta[n]$ and all its sub simplicial sets.

Proposition 1.3.2 *Let A be finite simplicial set and $X(0) \to X(1) \to \dots$ a sequence of simplicial sets. Then the natural map*

$$\varinjlim \mathcal{S}(A, X(i)) \to \mathcal{S}(A, \varinjlim X(i))$$

is a bijection.

Sketch Proof: Let us for simplicity assume that all the maps $X(i) \to X(i+1)$ are inclusions (that is all we needed in 1.2.2 anyhow). Then the natural map is perforce an inclusion, and we only need to show surjectivity.
    Let $f \colon A \to \varinjlim X(i) = \bigcup X(i) \in \mathcal{S}$. Every k-simplex in A must necessarily map to $X(i)_k$ for some i depending on the simplex, but since there are only finitely may non-degenerate simplices we may choose a specific i such that they all map to $X(i)$.
    If $x \in A_n$ is any simplex, there is a unique non-degenerate simplex y such that $x = \phi^* y$ for some surjective $\phi \in \Delta$, and so $f(x) = f(\phi^* y) = \phi^* f(y)$ must also be in $X(i)$ (since $X(i) \subseteq \bigcup X(j)$ is a simplicial map).   □

Remark 1. We have the small object argument for any $A \in \mathcal{S}$, provided the cardinality of the indexing of the colimit is sufficiently big.

1.4 Fibrations

There is **one** aspect which is more awkward in simplicial sets than in topological spaces, and we will try to explain this, and at the same time give a definition of fibrations.

Definition 1.4.1 Let n and k be integers satisfying $0 \le k \le n > 0$. The kth-horn

$$\Lambda^k[n] \subseteq \Delta[n]$$

be the sub-simplicial set generated by all faces of $\Delta[n]$ but the kth.

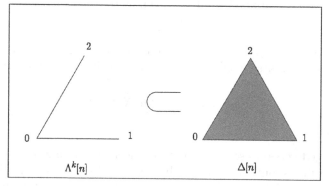

The inclusion of the horn, here illustrated with $n = 2$ and $k = 0$. The 0th horn is generated by all faces but the 0th, which is the face opposite to the 0th-vertex

Obviously, the inclusion of the horn into the standard simplex is a weak equivalence. If you realize, you get that $|\Lambda^k[n]| \subseteq |\Delta[n]|$ is a deformation retract:

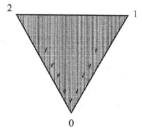

The inclusion $|\Lambda^0[2]| \subseteq |\Delta[2]|$ is a deformation retract by the retraction illustrated

However, there is no retraction before realizing. This is the only bad thing about simplicial sets, and unfortunately there is no pain-free cure.

However, there is a large class of simplicial sets – called *fibrant* – that do not have this problem. Fortunately, every simplicial set is weakly equivalent to a fibrant simplicial set:

Definition 1.4.2 A map $f\colon X \to Y \in \mathcal{S}$ is a *fibration* if it has the right lifting property with respect to all the inclusions $\Lambda^k[n] \subseteq \Delta[n]$ for $0 \le k \le n > 0$.
If $X \to *$ is a fibration, we say that X is *fibrant*.

Example 1.4.3   1. If Y is a topological space, then $\mathrm{sing}\, Y$ is fibrant.
2. If M is a simplicial group, then the underlying simplicial set UM is fibrant (essentially because you have inverses: see e.g., [6, I.3.4]).

Exercise 1.4.4 Prove that if $Y \in \mathcal{T}op$, then $\mathrm{sing}\, Y$ is fibrant.

Exercise 1.4.5 Note that a trivial fibration actually is a fibration.

Fact 1.4.6 One may show that being a trivial fibration is equivalent to being both a fibration and a weak equivalence.

Fact 1.4.7 Our definition of fibrations is equivalent to saying that a fibration is a map which has the right lifting property with respect to injections that are weak equivalences. To see this one has to show that inclusions that are weak equivalences can be built out of the filling of horns (more precisely, they are retracts of injections $X(0) \to \bigcup X(i)$ where each $X(i - 1) \to X(i)$ are pushouts of disjoint unions of $\Lambda^k[n] \subseteq \Delta[n]$'s).

Fact 1.4.8 Using the small object argument again we get that given any map $f\colon X \to Y$ there exists a (functorial) factorization

$$X \overset{\iota_f}{\underset{\sim}{\hookrightarrow}} Z_f \overset{\phi_f}{\longrightarrow} Y$$

of f into a inclusion that is a weak equivalence followed by a fibration.

Note 1.4.9 *In the literature fibrations of simplicial sets are often referred to as Kan fibrations and fibrant simplicial sets as Kan complexes.*

Proposition 1.4.10 *If $i\colon A \subseteq B$ and $f\colon X \to Y$ is a fibration, then the canonical map*

$$(i, p)_*\colon \underline{\mathcal{S}}(B, X) \to \underline{\mathcal{S}}(B, Y) \times_{\underline{\mathcal{S}}(A, Y)} \underline{\mathcal{S}}(A, X)$$

is a fibration. If either i or f are weak equivalences then so is $(i, p)_*$.

Sketch Proof: We only prove the case $A = \emptyset$ and $Y = *$ since this is all we need just now. In that case it reduces to showing that if X is fibrant, then so is $\underline{S}(B, X)$, and if in addition $X \to *$ is a weak equivalence, so is $\underline{S}(B, X) \to *$.

Consider a diagram of the sort

$$
\begin{array}{ccc}
K & \longrightarrow & \underline{S}(B, X) \\
\downarrow & & \downarrow \\
L & \longrightarrow & *
\end{array}
\quad .
$$

Asking about liftings is the same as asking whether the map

$$
\mathcal{S}(L, \underline{S}(B, X)) \to \mathcal{S}(K, \underline{S}(B, X))
$$

is surjective, which is the same as asking whether

$$
\mathcal{S}(B \wedge L, X) \to \mathcal{S}(B \wedge K, X)
$$

is surjective, which is the same as asking about liftings in the diagram below

$$
\begin{array}{ccc}
K \wedge B & \longrightarrow & X \\
\downarrow & & \downarrow \\
L \wedge B & \longrightarrow & *
\end{array}
\quad .
$$

But if $K \to L$ is injective or a weak equivalence, then so is $K \wedge B \to L \wedge B$ (injectivity is obvious, and weak equivalence may be seen by realizing). So since X is fibrant we get liftings for all injections $K \to L$ that are weak equivalences. If in addition $X \to *$ is a weak equivalence we get liftings for all inclusions by 1.4.6. $\qquad \square$

Proposition 1.4.11 *(the Whitehead theorem) If $X, Y \in \mathcal{S}$ are fibrant and $f \colon X \to Y$ is a weak equivalence, then f is a homotopy equivalence.*

Sketch Proof: By factorization, we may assume that f is either a trivial fibration or both an injection and a weak equivalence (that this works needs a slight checking). Assume first f is a trivial fibration (in which case it is not needed that X and Y are fibrant). Then there is a lifting $s \colon X \to Y$ in the diagram

$$
\begin{array}{ccc}
\emptyset & \longrightarrow & X \\
\downarrow & & \simeq \downarrow f \\
Y & = & Y
\end{array}
$$

and we must show that s is a homotopy inverse. By construction fs is the identity, but what about sf? Consider the diagram

$$X \coprod X \xrightarrow{\ sf + id_X\ } X$$

$$\downarrow \qquad\qquad \simeq \Big\downarrow f.$$

$$X \times \Delta[1] \xrightarrow{\ f\,pr_X\ } Y$$

The left vertical map is the inclusion induced by $\partial\Delta[1] \subseteq \Delta[1]$, and since f is a trivial fibration we get a lifting of the homotopy, giving us the desired homotopy $X \times \Delta[1] \to X$ from sf to the identity.

On the other hand, assume that f is an inclusion and a weak equivalence. Then the diagram

$$X =\!=\!=\!= X$$

$$f\Big\downarrow \qquad\qquad \Big\downarrow$$

$$Y \longrightarrow *$$

has a lifting $p\colon Y \to X$ since X is fibrant. We must show that fp is homotopic to the identity. Consider the diagram

$$Y \coprod_X (X \times \Delta[1]) \coprod_X Y \xrightarrow{\ fp + f\,pr_X + id_Y\ } Y$$

$$\downarrow \qquad\qquad\qquad\qquad \Big\downarrow .$$

$$Y \times \Delta[1] \qquad\qquad \longrightarrow \qquad *$$

Again we get a lifting (since Y is fibrant and the left vertical map is an injection and a weak equivalence), and the desired homotopy is established. $\qquad\square$

Example 1.4.12 Check that the above proof using factorizations give a proof in general.

2 Combinatorial Homotopy Groups

We are now in a position to define homotopy groups for simplicial sets entirely within $\mathcal{S}$ without reference to topological spaces. This can be done in one of many ways, the first we present is not very constructive since it involves the small object argument. The second is leaner and involves Kan's particular "fibrant replacement functor Ex^∞.

2.1 Homotopies and Fibrant Objects

Consider the path relation between vertices discussed in 3.0.2. There we mentioned that generally symmetry and transitivity was was a problem. This is not so for fibrant X.

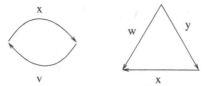

The symmetry and reflexivity of the path relation in fibrant objects

Exercise 2.1.1 Prove that in fibrant simplicial sets the path relation is an equivalence relation.

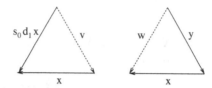

The idea of the proof: give 1-simplices x and y with $d_0 x = d_1 y$, find 2-simplices in X with the desired boundary

This opens for the definition:

Definition 2.1.2 If $X \in \mathcal{S}_*$ is fibrant, then

$$\pi_q X = \pi_0 \underline{\mathcal{S}}_*(S^q, X) \ .$$

Exercise 2.1.3 Prove that $\pi_1 X$ is a group. If you are really industrious you may want to show directly that $\pi_q X$ is an abelian group for $q > 1$.

Lemma 2.1.4 *If $X \in \mathcal{S}_*$ is fibrant, then there is a natural isomorphism $\pi_q X \cong \pi_* |X|$.*

Sketch Proof: Since X is fibrant we have by 1.4.11 that the weak equivalence $X \to \mathrm{sing}\,|X|$ is a homotopy equivalence, and so that $\underline{\mathcal{S}}_*(S^q, X) \to \underline{\mathcal{S}}_*(S^q, \mathrm{sing}\,|X|) \cong \mathrm{sing}\,(|X|^{|S^q|})$ is a homotopy equivalence. Hence they have the same π_0, giving the desired result.   □

So if $X \in \mathcal{S}_*$ we can choose any (pointed) weak equivalence $X \xrightarrow{\sim} Y$ with Y fibrant and define the homotopy groups of X to be those of Y. Of course, we want the choice of Y to be functorial, and one way is to use the functorial factorization of $X \to *$ into a weak equivalence that is an inclusion followed by a fibration. Another way is to simply let $Y = \mathrm{sing}\,|X|$.

Exercise 2.1.5 Let $f\colon X \to Y$ be a fibration and $y \in Y_0$, and define the *fiber* over y to be the simplicial set F gotten by the pullback

$$
\begin{array}{ccc}
F & \longrightarrow & X \\
\downarrow & & f\downarrow \\
* & \xrightarrow{\ y\ } & Y
\end{array}
$$

(i.e., $F \cong f^{-1}(y)$). Prove that F is fibrant. In addition, if X and Y also are fibrant and X and F are pointed in an $x \in f^{-1}(y)$, prove the long exact sequence

$$\cdots \to \pi_{n+1}Y \to \pi_n F \to \pi_n X \to \pi_n Y \to \cdots \to \pi_0 Y \ .$$

Exercise 2.1.6 Use the fact that the result of the last exercise is true for arbitrary fibrations to prove that if $f\colon X \to Y$ is a trivial fibration, then f is a weak equivalence.

Exercise 2.1.7 Prove that if G is a simplicial group, then the path relation is an equivalence relation [in fact: G is fibrant]. Prove that $\pi_0 G$ is a group. Prove that $\mathcal{S}_*(S^q, G)$ is a simplicial group.

2.1.8 Subdivisions and Kan's Ex^∞

For completeness we present Kan's Ex^∞ which is a particularly compact "fibrant replacement functor" related to the small object argument. We refer to [6, III4] for proofs of the claims presented in this section.

Consider the category Inj with objects the finite subsets of $\mathbf{N}$, and with morphisms the inclusions (this is equivalent to the subcategory of Δ of all injective maps, but the combinatorics becomes easier this way).

The *over category* Inj_n is the category whose objects are subsets $S \subseteq \{0, 1, \ldots, n\}$ and where there is a single morphism from $T \subseteq \{0, 1, \ldots, n\}$ to $S \subseteq \{0, 1, \ldots, n\}$ if $T \subseteq S \subseteq \{0, 1, \ldots, n\}$

For every $\phi\colon [n] \to [m] \in \Delta$ we get a functor $\phi_*\colon \mathrm{Inj}_n \to \mathrm{Inj}_m$ by sending $S \subseteq \{0, 1, \ldots, n\}$ to $\phi(S) \subseteq \{0, 1, \ldots, m\}$ making

$$[n] \to \mathrm{Inj}_n$$

a cosimplicial category. The functor $\mathrm{Inj}_n \to [n]$ sending $\{i_0, \ldots, i_q\} \subseteq \{0, 1, \ldots, n\}$ to $i_q \in [n]$ becomes a map of cosimplicial categories when $[n]$ varies.

For any simplicial set X Kan then defines

$$\mathrm{Ex}(X) = \{[q] \mapsto \mathcal{S}(N(\mathrm{Inj}_q), X)\}$$

This is a simplicial set, and $N(\mathrm{Inj}_q) \to N[q] = \Delta[q]$, defines an inclusion $X \subseteq \mathrm{Ex}(X)$. Set

$$\mathrm{Ex}^\infty X = \varinjlim_k \mathrm{Ex}^{(k)}(X) \ .$$

Fact 2.1.9 Kan's subdivision functor Ex^∞ is a fibrant replacement functor, i.e.,

1. The inclusion $X \subseteq \mathrm{Ex}^\infty X$ is a weak equivalence (pointed if X is pointed).
2. $\mathrm{Ex}^\infty X$ is fibrant.

Kan then defines the homotopy groups of X without reference to topological spaces via

$$\pi_q X = \pi_0 \underline{\mathcal{S}}_*(S^q, \mathit{Ex}^\infty X) .$$

III

Model Categories

We are interested in homotopy theory in a variety of categories, and we want to compare these. There is an efficient machinery due to Quillen, which encodes this structure. We have used this language in our discussion of simplicial sets. In addition to weak equivalences (which is all that is needed to form the homotopy category) we have fibrations and cofibrations satisfying certain axioms. This structure ensures that the homotopy category actually exists, but more importantly it encodes the deeper homotopical structures, making a large class of arguments formal. It also makes comparison between different homotopical structures more transparent.

In particular, we are interested in functor categories. If I is a small category and $\mathcal{M}$ is a category "in which we know how to do homotopy theory", how can we do homotopy theory in the category of functors from I to $\mathcal{M}$? This question does not have a unique answer (which is a good thing since that different answers are serviceable in different situations), but there is still much we can say.

0.1 Liftings

As discussed at the beginning of the previous chapter, the formal side of homotopy theory is all about liftings.

Definition 0.1.1 Let $\mathcal{M}$ be a category, and $i\colon A \to B$ and $p\colon X \to Y$ be two maps in $\mathcal{M}$. We say that p has the *right lifting property* with respect to i (and that i has the *left lifting property* with respect to p) if for all commutative diagrams

$$\begin{array}{ccc} A & \longrightarrow & X \\ i\downarrow & & \downarrow p \\ B & \longrightarrow & Y \end{array}$$

in $\mathcal{M}$, there is a map $s\colon B \to X \in \mathcal{M}$ making the following diagram commutative

1 The Axioms

A model category is a category equipped with (more than) enough structure to do homotopy theory. The axioms are good in the sense that they apply to most of the situations we could imagine desirable, and still they are strong enough to carry important information.

Given the complexity of the axioms, it is perhaps surprising that they have not changed significantly during the 35 years that have passed since Quillen first proposed them. It is perhaps even more surprising that the modifications that have been suggested all have tended to make the axioms even more restrictive.

We give the version most common these days, following [4], see also [8]. Hovey [9] insists on a choice instead of merely existence in the factorization axiom $\mathcal{M}5$ below:

Definition 1.0.1 A *model category* is a category $\mathcal{M}$ together with three classes of maps cof$\mathcal{M}$ (the *cofibrations*), fib$\mathcal{M}$ (the *fibrations*) and w$\mathcal{M}$ (the *weak equivalences*) satisfying the following axioms:

$\mathcal{M}1$: (Limit axiom) The category $\mathcal{M}$ is (co)complete (i.e., has all small (co)limits).

$\mathcal{M}2$: (Two out of three axiom) If
$$X \xrightarrow{\ g\ } Y \xrightarrow{\ f\ } Z \in \mathcal{M}$$
and two of f, g and fg are weak equivalences, then so is the third.

$\mathcal{M}3$: (Retract axiom) If the map $g \in \mathcal{M}$ is a retract of $h \in \mathcal{M}$ and h is in cof$\mathcal{M}$, fib$\mathcal{M}$ or in w$\mathcal{M}$, then so is g.

$\mathcal{M}4$: (Lifting axiom) The cofibrations have the left lifting property with respect to the maps in fib$\mathcal{M} \cap$w$\mathcal{M}$ (the *trivial fibrations*) and the fibrations have the right lifting property with respect to the maps in cof$\mathcal{M} \cap$w$\mathcal{M}$ (the *trivial cofibrations*).

$\mathcal{M}5$: (Factorization axiom) If $g\colon X \to Y \in M$ there exist functorial factorizations

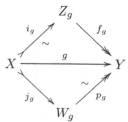

where i_g is a trivial cofibration, f_g is a fibration, j_g is a cofibration and p_g is a trivial fibration.

Note 1.0.2 *Note that a model category always have initial and final objects. An object for which the unique map from the initial object is a cofibration is said to be cofibrant, and – dually – an object for which the unique map to the final object is a fibration is said to be fibrant.*

Theorem 1.0.3 *The categories S, S_*, A, Top, Top_* and Spt "are" model categories.*

By saying that they "are" model categories we mean that the categories have choices of subcategories which make them into model categories, and the weak equivalences are what you think they are.

To be a bit more precise:

1. (S and S_*) The injections in S and S_* are the cofibrations (so all objects are cofibrant) and the weak equivalences and fibrations were discussed in the previous chapter.
2. (Top and Top_*) The fibrations are the maps f such that sing f are fibrations (a.k.a. "Serre fibrations"), the weak equivalences are those inducing isomorphisms on π_* and the cofibrations are those that have the left lifting property with respect to the trivial fibrations. All objects are fibrant.
3. (A) A map $f \in A = sAb$ is a weak equivalence (resp. fibration) if it is so when forgetting down to S. The cofibrations are those with the left lifting property with respect to the trivial fibrations. All objects are fibrant (a map $X \to Y$ is a fibration if the map $X \to Y \times_{\pi_0 Y} \pi_0 X$ is a surjection, and the cofibration condition is very closely associated with free generation of non-degenerate elements).
4. (Spt) A map in Spt is a weak equivalence if it is a stable equivalence. The fibrations and cofibrations will be discussed in Sect. 5.

1.1 Simple Consequences

The axioms are stated in a form intended to make them easy to check. In practice, you often know that the category in question is a model category, and then it is better to know some of the implications.

Exercise 1.1.1 (see e.g., [8, 7.2]).

1. Let $g \colon X \to Y \in M$ be factored as $g = pi$ where p has the right lifting property with respect to g. Show that g is a retract of i.
2. A map is in $\mathrm{cof} M$ (resp. $\mathrm{cof} M \cap \mathrm{w} M$) if **and only if** it has the left lifting property with respect to the maps in $\mathrm{fib} M \cap \mathrm{w} M$ (resp. $\mathrm{fib} M$). A map is in $\mathrm{fib} M$ (resp. $\mathrm{fib} M \cap \mathrm{w} M$) if **and only if** it has the right lifting property with respect to the maps in $\mathrm{cof} M \cap \mathrm{w} M$ (resp. $\mathrm{cof} M$).
3. Two of $\mathrm{cof} M$, $\mathrm{fib} M$ and $\mathrm{w} M$ determine the third.

4. The classes of maps $\mathrm{cof}\mathcal{M}$, $\mathrm{fib}\mathcal{M}$ and $\mathrm{w}\mathcal{M}$ form subcategories of $\mathcal{M}$ containing all objects (and all isomorphisms).

5. If

$$
\begin{array}{ccc}
A & \xrightarrow{\ i\ } & C \\
{\scriptstyle g}\downarrow & & \downarrow{\scriptstyle f} \\
B & \xrightarrow{\ j\ } & D
\end{array}
$$

is a commutative square in $\mathcal{M}$ we have that

    a) if the square is a pushout square and i is a (trivial) cofibration, then so is j.

    b) if the square is a pullback square and f is a (trivial) fibration, then so is g.

Note 1.1.2 *On terminology: In [21] Quillen only required finite (co)limits, used a weaker version of the lifting Axiom 1.0.14 (reserving the term closed model category for the current version) and did not require that the factorizations in the factorization Axiom 1.0.15 had to be functorial.*

He used the terms trivial (co)fibration in [21] and [19], but switched to acyclic (co)fibration in [20] (where he states that '"acyclic" is much preferable to the term "trivial"'). You will see both forms in the literature, and there are good arguments against both.

One of the most useful technical tools in model categories has become known as *Ken Brown's lemma* (earlier each new generation of students reading Quillen's works had to rediscover this lemma for themselves, but nowadays it is on the standard repertoire). We include the proof since it is a good example of the axioms at play.

Lemma 1.1.3 *Let $\mathcal{M}$ be a model category. If $f\colon A \to B$ is a map between cofibrant objects, then there exists a (functorial) diagram*

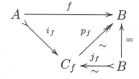

where i_f is a cofibration, p_f is a trivial fibration and j_f a trivial cofibration.

    Dually, if $f\colon A \leftarrow B$ is a map between fibrant objects, then there exists a (functorial) diagram

$$
\begin{array}{ccc}
A & \xleftarrow{\ f\ } & B \\
{\scriptstyle P_f}\nwarrow \quad {\scriptstyle I_f}\nearrow & & \Vert{\scriptstyle =} \\
Z_f & \xrightarrow[\sim]{\ Q_f\ } & B
\end{array}
$$

where I_f is a trivial cofibration, P_f is a fibration and Q_f a trivial fibration.

Sketch Proof: We prove only the first part. Using the factorization axiom $\mathcal{M}5$, factor the map

$$f + id_B \colon A \coprod B \to B$$

into a cofibration followed by a trivial fibration

$$A \coprod B \rightarrowtail C_f \xrightarrow{\;p_f\;}_{\sim} B$$

Since A and B are cofibrant, and cofibrations are closed under pushouts by 1.1.1.5, both the canonical maps $A \to A \coprod B \leftarrow B$ are cofibrations. Noting that the diagram

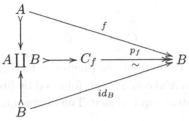

commutes, we are done (use that compositions of cofibrations are cofibrations (1.1.1.4) and the two out of three axiom $\mathcal{M}2$). $\qquad\square$

As an immediate consequence we have:

Corollary 1.1.4 *A functor $F \colon \mathcal{M} \to \mathcal{N}$ between model categories that preserve trivial cofibrations sends weak equivalences between cofibrant objects to weak equivalences.*

1.2 Proper Model Categories

Definition 1.2.1 Let $\mathcal{M}$ be a model category. We say that $\mathcal{M}$ is *left proper* if for all pushout diagrams

$$
\begin{array}{ccc}
A & \xrightarrow{\;f\;}_{\sim} & B \\
\downarrow{\scriptstyle i} & & \downarrow \\
C & \xrightarrow{\;g\;} & D
\end{array}
$$

with i a cofibration and f a weak equivalence, g is a weak equivalence.

Dually, $\mathcal{M}$ is *right proper* if for all pullback diagrams

$$
\begin{array}{ccc}
A & \xrightarrow{\;g\;} & B \\
\downarrow & & \downarrow{\scriptstyle p} \\
C & \xrightarrow{\;f\;}_{\sim} & D
\end{array}
$$

with p a fibration and f a weak equivalence, g is a weak equivalence.

If $\mathcal{M}$ is both left and right proper, we say that $\mathcal{M}$ is *proper*.

Example 1.2.2 The model category structures on $\mathcal{S}$, $\mathcal{S}_*$, $\mathcal{A}$, $\mathcal{T}op$, $\mathcal{T}op_*$ and $\mathcal{S}pt$ are all proper (see e.g., [8, 11.1], which also implies the case $\mathcal{A} = s\mathcal{A}b$. For spectra, see [1]).

For future reference we include the following definition

Definition 1.2.3 In a model category a (commuting) square

$$
\begin{array}{ccc}
A & \longrightarrow & B \\
\downarrow & & \downarrow \\
C & \longrightarrow & D
\end{array}
$$

is a *homotopy pullback square* if for the functorial factorizations $B \overset{\sim}{\rightarrowtail} X \twoheadrightarrow D$ and $C \overset{\sim}{\rightarrowtail} Y \twoheadrightarrow D$ (into trivial cofibrations followed by fibrations) the canonical map $A \to X \times_D Y$ is a weak equivalence. The *homotopy pullback* is the object $X \times_D Y$.

The good thing about homotopy pullbacks in (right) proper model categories is that they are homotopy invariant, and if the square in the definition were already a (categorical) pullback diagram and $B \to D$ a fibration, then it would be homotopy cartesian, see [8].

1.3 Quillen Functors

The morphisms in "the category of model categories" is not what you first come to think about (namely a functor preserving all the structure: this is much too rigid).

Definition 1.3.1 Let $\mathcal{M}$ and $\mathcal{N}$ be model categories. A *(left) Quillen functor* from $\mathcal{M}$ to $\mathcal{N}$ is a functor $F\colon \mathcal{M} \to \mathcal{N}$ such that

- F has a right adjoint
- F preserves cofibrations and trivial cofibrations.

A Quillen functor F with right adjoint G is called a *Quillen equivalence* if for all cofibrant $A \in \mathcal{M}$ and fibrant $X \in \mathcal{N}$ the bijection

$$\mathcal{M}(A, GX) \cong \mathcal{N}(FA, X)$$

restricts to a bijection

$$w\mathcal{M}(A, GX) \cong w\mathcal{N}(FA, X)$$

Example 1.3.2 1. The realization $\mathcal{S} \to \mathcal{T}op$ is a Quillen equivalence (see below)
  2. The adding base point gives a Quillen functor $(-)_+ \colon \mathcal{S} \to \mathcal{S}_*$.
  3. The free functor $\tilde{\mathbf{Z}} \colon \mathcal{S}_* \to \mathcal{A}$ is a Quillen functor.
  4. The suspension spectrum gives a Quillen functor $\Sigma^\infty \colon \mathcal{S}_* \to \mathcal{S}pt$.

Fact 1.3.3 If both a Quillen functor F and its right adjoint G preserve weak equivalences, F is a Quillen equivalence if and only if for all cofibrant and fibrant objects $A \in \mathcal{M}$ and cofibrant and fibrant objects $X \in \mathcal{N}$ the adjunction maps

$$A \to GFA \text{ and } FGX \to X$$

are weak equivalences.

From this we see

Theorem 1.3.4 *The realization $\mathcal{S} \to \mathcal{T}op$ is a Quillen equivalence.*

Exercise 1.3.5 Prove that the right adjoint of a (left) Quillen functor preserves fibrations and trivial fibrations.

2 Functor Categories: The Projective Structure

Definition 2.0.1 Let $\mathcal{C}$ be a small category, and consider the category $\mathcal{S}^{\mathcal{C}}$ of functors $\mathcal{C} \to \mathcal{S}$. Call a natural transformation $X \to Y \in \mathcal{S}^{\mathcal{C}}$ a *pointwise weak equivalence* (resp. *pointwise fibration*) if for every $c \in \mathcal{C}$ the map $X(c) \to Y(c) \in \mathcal{S}$ is a weak equivalence (resp. fibration).
  A *cellular inclusion* is a composite

$$X(0) \to X(1) \to \cdots \to \varinjlim_{i \in \mathbf{N}} X(i)$$

where each $X(i) \to X(i+1)$ is the pushout along a disjoint union of maps

$$\partial\Delta[n] \times \mathcal{C}(c, -) \subseteq \Delta[n] \times \mathcal{C}(c, -)$$

for $c \in \mathcal{C}$ and $n \geq 0$.

Theorem 2.0.2 *(see e.g., [2, p. 314])* *The above gives a model category structure, called the projective structure, on $\mathcal{S}^{\mathcal{C}}$ in which a map is a cofibration iff it is a retract of a cellular inclusion.*

Sketch Proof: One proof follows the line of the same claim for $\mathcal{S}$ (which relies on facts about the realization functor as discussed in the previous chapter). Alternatively one can use a lifting lemma approach as in [4]. $\square$

Note 2.0.3 *For the application to motivic homotopy theory (see Chap. IV), this is not the model structure on the simplicial presheaves $Sm/S^{op} \to S_*$ usually considered. However, it is quite useful e.g., when considering spectra through motivic functors, partially because it is much "smaller" than the one considered by Jardine, Morel and Voevodsky (in which the cofibrations are the injections).*

3 Cofibrantly Generated Model Categories

There was one particularly cool thing about simplicial sets which we employed many times: that the "building blocks" for (trivial) cofibrations were easy to cope with.

Definition 3.0.1 Let I be a set of the morphisms in a cocomplete category $\mathcal{M}$. We let

1. $I-$inj be the set of morphisms in $\mathcal{M}$ having the right lifting property with respect to the maps in I,
2. $I-$cof be the set of morphisms in $\mathcal{M}$ having the left lifting property with respect to the maps in $I-$inj. The elements of $I-$cof are referred to as *I-cofibrations*.

Previously when we discussed the small object argument, it referred to colimits indexed over the natural numbers. In what follows we will have to allow colimits over larger ordinals (satisfying the property that for all limit ordinals γ the map $\lim_{\overrightarrow{\beta<\gamma}} X(\beta) \to X(\gamma)$ is an isomorphism), but we refer to the literature for making this transfinite mumbojumbo precise e.g., [9, 2.1].

That a set I of maps permits the small object means that the domains of the maps are small with respect to (transfinite) compositions of pushouts of coproducts of maps in I (a.k.a. I-cell).

Definition 3.0.2 A *cofibrantly generated model category* is a model category $\mathcal{M}$ in which there exists

1. a set I of cofibrations which permits the small object argument and for which the cofibrations and I-cofibrations coincide.
2. a set J of trivial cofibrations which permits the small object argument and for which the trivial cofibrations and J-cofibrations coincide.

Example 3.0.3 All the model categories we have discussed so far are cofibrantly generated.

Lemma 3.0.4 *(Recognition principle). Let $\mathcal{M}$ be a (co)complete category with a subcategory $w\mathcal{M} \subseteq \mathcal{M}$ which is closed under retracts and satisfies the two-out-of-three axiom. Call the morphisms in $w\mathcal{M}$ weak equivalences.*

Let I and J be sets of maps in $\mathcal{M}$. Call the maps in $I-$cof (resp. $J-$inj) cofibrations (resp. fibrations) and the maps in $J-$cof (resp. $I-$inj) trivial cofibrations (resp. fibrations).

Assume that

1. *I and J permit the small object argument;*
2. *the trivial cofibrations are both cofibrations and weak equivalences and the trivial fibrations are both fibrations and weak equivalences;*
3. *and either all maps that are both cofibrations and weak equivalences are trivial cofibrations, or all maps that are both fibrations and weak equivalences are trivial fibrations.*

Then this structure defines a cofibrantly generated model structure on $\mathcal{M}$.

Sketch Proof: The first Axioms 1–3 follow automatically. The small object argument then gives 5. The third condition then implies one of the halves of 4 while the other follow from 5 and a retraction argument as in 1.1.1.1 (showing e.g. that all maps that are both weak equivalences and cofibrations necessarily are retracts of trivial cofibrations, and hence trivial cofibrations) $\square$

Let $\mathcal{C}$ be a small category, and consider the category $\mathcal{S}^{\mathcal{C}}$ of functors $\mathcal{C} \to \mathcal{S}$. Call a natural transformation $X \to Y \in \mathcal{S}^{\mathcal{C}}$ a *pointwise weak equivalence* (resp. *pointwise cofibration*) if for every $c \in \mathcal{C}$ the map $X(c) \to Y(c) \in \mathcal{S}$ is a weak equivalence (resp. cofibration).

Theorem 3.0.5 *(Heller, see [4]) The structure above gives rise to a model category structure on $\mathcal{S}^{\mathcal{C}}$ which is Quillen equivalent to the projective structure 2.0.2 through the identity map (from the projective to the current).*

Lemma 3.0.6 *(Lifting lemma). Let $\mathcal{M}$ be a cofibrantly generated model category, and let I (and J) be sets of generating (trivial) cofibrations. Let $\mathcal{C}$ be a (co)complete category and*

$$\mathcal{C} \underset{R}{\overset{L}{\leftrightarrows}} \mathcal{M}$$

be a pair of adjoint functors such that LI and LJ admit the small object argument and such that Rf is a weak equivalence for every $f \in LJ-$cof.

Then $\mathcal{C}$ has a cofibrantly generated model structure with where a map $f \in \mathcal{C}$ is a weak equivalence (resp. fibration) if $Rf \in \mathcal{M}$ is. Furthermore the sets LI and LJ are generating (trivial) cofibrations and L is a Quillen functor.

Sketch Proof: Use the recognition Principle 3.0.4. $\square$

Exercise 3.0.7 Use the lifting lemma to prove that the structure on $\mathcal{A}$ given in the explanation following Theorem 1.0.3 actually is a model structure.

4 Simplicial Model Categories

It is often the case that the set of maps assemble to a function *space* (or simplicial set), and that the model structure conforms nicely with the homotpy theory of these function spaces. In these cases many things become somewhat more transparent.

Definition 4.0.1 An *$\mathcal{S}$-category* $\mathcal{C}$ is a class of "objects" $ob\mathcal{C}$ together with a simplicial set
$$\underline{\mathcal{C}}(A, X) \in \mathcal{S}$$
for each pair of objects $A, X \in ob\mathcal{C}$ with a unital and associative composition (the usual axioms for a category, just allowing morphism sets to be simplicial sets).

Note 4.0.2 *An $\mathcal{S}$-category $\mathcal{C}$ has an underlying category $\mathcal{C}$ by letting*
$$\mathcal{C}(A, X) = \underline{\mathcal{C}}(A, X)_0 \ .$$

Note that it takes no effort to replace $\mathcal{S}$ by similar animals like $\mathcal{S}_*$ or $\mathcal{A}$, giving rise to parallel theories.

Exercise 4.0.3 Give a definition of an $\mathcal{S}$-functor, which you think will be useful.

Definition 4.0.4 A $\mathcal{S}$-category is *tensored and cotensored* if for all $X \in \mathcal{C}$ the functor
$$Y \mapsto \underline{\mathcal{C}}(X, Y)$$
has a left adjoint $K \mapsto X \otimes X$, and if $K \mapsto X \otimes K$ has a right adjoint (everything natural).

Definition 4.0.5 A *simplicial model category* is a tensored and cotensored $\mathcal{S}$-category $\mathcal{M}$ with a model category structure on the underlying category, such that for all cofibrations $i \colon A \rightarrowtail B$ and fibrations $p \colon X \twoheadrightarrow Y$, the canonical map
$$(i, p)_* \colon \underline{\mathcal{M}}(B, X) \to \underline{\mathcal{M}}(B, Y) \times_{\underline{\mathcal{M}}(A,Y)} \underline{\mathcal{M}}(A, X) \in \mathcal{S}$$
is a fibration, and that furthermore, if in addition either i or p are weak equivalences, then so is $(i, p)_*$.

Note 4.0.6 *Quillen referred to the "tensored and cotensored" part as axiom SM0 and to the condition on the map $(i, p)_*$ as axiom SM7. The terminology "simplicial model category" is obviously unfortunate, as one would think that it referred to a functor from Δ^{op} to some category of model categories, but the terminology is well established.*

Note 4.0.7 *In simplicial model categories we have the notion of (simplicial) homotopy at our disposal. This means that we have means of detecting weak equivalences at the function space level.*

Proposition 4.0.8 *(see e.g., [8, 10.5.1]) Let $\mathcal{M}$ be a simplicial model category. A map $f: Y \to Z$ is a weak equivalence if either of the following conditions are satisfied:*

1. *For every fibrant object $X \in \mathcal{M}$ the induced map*

$$\underline{\mathcal{M}}(Z, X) \to \underline{\mathcal{M}}(Y, X) \in \mathcal{S}$$

   is a weak equivalence.
2. *For every cofibrant object $A \in \mathcal{M}$ the induced map*

$$\underline{\mathcal{M}}(A, Y) \to \underline{\mathcal{M}}(A, Z) \in \mathcal{S}$$

   is a weak equivalence.

In the case where both Y and Z are cofibrant the first condition is necessary and sufficient. Likewise with fibrant vs. the latter condition.

5 Spectra

As we discussed in I.6, in topology spectra are the necessary device for applying homotopy theory on the collection of cohomology theories. This is also true in motivic homotopy theory; in order to study cohomology theories like motivic cohomology and algebraic K-theory we introduce spectra. An interesting thing is what replaces the circle: spectra study "stable" phenomena, i.e., smashing with the circle shall be an equivalence of homotopy categories. In order to study this effectively we present a model structure on the category of spectra making this true.

5.1 Pointwise Structure

This structure is just a technical device, preparing the ground for the stable model structure.

Definition 5.1.1 A pointwise weak equivalence (resp. fibration) of spectra is a map $X \to Y \in \mathcal{S}pt$ such that for every $n \geq 0$ the map $X^n \to Y^n$ is a weak equivalence (resp. fibration) in $\mathcal{S}_*$. A cofibration is a map with the left lifting property with respect to the maps that are both pointwise equivalences and pointwise fibrations.

Exercise 5.1.2 The pointwise structure is a model category.

Note 5.1.3 *As a matter of fact, Spt is isomorphic to a functor category $[\mathcal{S}_{S^1}, \mathcal{S}_*]$ if one allows the enrichment in $\mathcal{S}_*$ to play a rôle (see e.g., [8, 9.6.2]). The model structure follows from the simplicial analog of the projective structure discussion above. The cofibrations are of the form $R(-) \wedge \partial\Delta[n]_+ \to R(-) \wedge \Delta[n]_+$ where R runs over the representable functors, and likewise for the trivial cofibrations.*

To be explicit, the objects of $\mathcal{S}_{S^1}$ are the S^n's for $n \geq 0$, and

$$\underline{\mathcal{S}_{S^1}}(S^n, S^{n+k}) = \begin{cases} S^k & \text{if } k \geq 0 \\ * & \text{otherwise} \end{cases}$$

(the composition is obtained from the natural isomorphisms $S^k \wedge S^l \cong S^{k+l}$).

Both symmetric spectra [11] and simplicial functors [15] can be treated in the same manner, making comparison easier.

5.2 Stable Structure

Definition 5.2.1 A *stable fibration* is a map in Spt with the right lifting property with respect to all cofibrations that are stable equivalences I.6.3.3. The *stable structure* on Spt consists of the stable equivalences, the stable fibrations and the cofibrations.

Proposition 5.2.2 *(see [1]) The stable structure defines a model category structure on Spt.*

Note 5.2.3 *[15] This model structure is cofibrantly generated: the generating cofibrations are the same as for the pointwise structure. The generating trivial cofibrations are given as follows using the notation of 5.1.3: consider the evaluation map*

$$ev_n \colon \mathcal{S}_{S^1}(S^1 \wedge S^n, -) \wedge S^1 \cong \mathcal{S}_{S^1}(S^1, \mathcal{S}_{S^1}(S^n, -)) \wedge S^1 \to \mathcal{S}_{S^1}(S^n, -)$$

and let s_n be the inclusion of $\mathcal{S}_{S^1}(S^1 \wedge S^n, -) \wedge S^1$ into the mapping cylinder of ev_n. Let the generating trivial cofibrations be the generating trivial cofibrations from the pointwise structure plus all the s_ns.

Let

$$(QX)^n = \varinjlim_i \operatorname{sing} \Omega^i |X^{i+n}|.$$

Note that this defines a spectrum, and there is a canonical map $X \to QX$.

Exercise 5.2.4 Check the details about QX.

This construction is occasionally referred to as "spectrification". The important property of QX is that it is an Ω-spectrum, i.e., the adjoint $(QX)^n \to \Omega(QX)^{n+1}$ is a weak equivalence.

Fact 5.2.5 (see [1]) A map $X \to Y \in \mathcal{S}pt$ is a stable fibration if and only if it is a pointwise fibration such that for all $n \geq 0$ the square

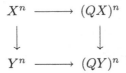

is a homotopy pullback 1.2.3 (which means in this situation that the canonical map from X^n to the pullback of the rest of the diagram is supposed to be a weak equivalence).

IV

Motivic Spaces and Spectra

As our last application of the machinery discussed in these talks, we come to an approach to the main topic of this summer school: motivic spaces and spectra.

We first discuss the spaces and then the spectra (both very briefly). I emphasize that this is just one approach, and that there are many others giving roughly the same result. The aim of this chapter is to show how the techniques we have discussed give a streamlined construction, and leave applications and algebro-geometric content to the other lectures. See [22] for a more thorough discussion and [13] for background from algebraic geometry.

1 Motivic Spaces

Let S be a Noetherian scheme and let Sm/S be the category of smooth scheme of finite type over S. Let $\mathcal{M}_S$ be the category of pointed simplicial presheaves on Sm/S, i.e., functors $(Sm/S)^{\mathrm{op}} \to \mathcal{S}_*$. This category is referred to as *the (motivic) category of spaces*, and is the basic object of study in this section.

We will provide $\mathcal{M}_S = \mathcal{S}_*^{(Sm/S)^{\mathrm{op}}}$ with several model category structures. Eventually we will have one whose homotopy category is worthy of the name *the unstable motivic homotopy category*. This is but one of many categories that can bear this name, but they are all Quillen equivalent, and the one we are going to present requires the least machinery.

We will study this category and also its functors briefly, before we go on to stabilize and so have motivic spectra in the next section.

Theorem III.2.0.2 provides a model structure on $\mathcal{M}_S = \mathcal{S}_*^{(Sm/S)^{\mathrm{op}}}$, called the projective structure. Furthermore (like any other functor category into $\mathcal{S}_*$), $\mathcal{M}_S$ has a smash product: if $X, Y \in \mathcal{M}_S$, then

$$(X \wedge Y)(U) = X(U) \wedge Y(U)$$

and an internal morphism object

$$\underline{\mathcal{M}}_S(X, Y) \in \mathcal{M}_S$$

right adjoint to the smash, and this satisfies the property corresponding to III.4.0.5, just that the morphism objects stays within $\mathcal{M}_S$ (they are *motivic spaces*, not just spaces). We say that $\mathcal{M}_S$ is a *monoidal model category*.

However, the projective structure is definitely not the structure we are interested in on $\mathcal{M}_S$. Firstly we have to take into account some topology, and secondly we will want the affine line to be contractible. This can be fixed as follows. The ideology is that we specify the fibrant objects as those objects having some desired property – at least up to homotopy – and model weak equivalences and cofibrations on them.

The topology we will consider here is the so-called *Nisnevich topology* (see [13] for this and other notions pertaining to algebraic geometry such as *smooth, étale and reduced*).

Definition 1.0.1 A space $X \in \mathcal{M}_S$ is *locally fibrant* if

1. $X(\emptyset) = *$,
2. for all $U \in Sm/S$ the pointed simplicial set $X(U)$ is fibrant, and
3. for all pullback squares

$$
\begin{array}{ccc}
P & \longrightarrow & Y \\
\downarrow & & \downarrow{\phi} \\
U & \xrightarrow{\ i\ } & Z
\end{array}
\quad \in Sm/S
$$

where ϕ is étale, i is an open imbedding and $\phi(Z - U) \to Z - U$ is an isomorphism in the reduced structure, the induced square

$$
\begin{array}{ccc}
X(P) & \longleftarrow & X(Y) \\
\uparrow & & \uparrow \\
X(U) & \longleftarrow & X(Z)
\end{array}
\quad \in \mathcal{S}_*
$$

is a homotopy pullback square (see III.1.2.3).

So, X is locally fibrant if it is a "Nisnevich sheaf up to homotopy".

For any $X \in \mathcal{M}_S$ let $* \rightarrowtail KX \xrightarrow{\sim} X$ be a functorial factorization (in the projective structure) of the canonical map $* \to X \in \mathcal{M}_S$.

Definition 1.0.2 A map $X \to Y \in \mathcal{M}_S$ is a *local equivalence* if for all locally fibrant Z the map

$$\underline{\mathcal{M}}_S(KY, Z) \to \underline{\mathcal{M}}_S(KX, Z)$$

is a pointwise equivalence. A map is a *local cofibration* if it is a projective cofibration, and a *local fibration* if it has the right lifting property with respect to to the local trivial cofibrations.

Note 1.0.3 *The local structure is a model category, and on locally fibrant objects local weak equivalences are detected pointwise.*

1.1 The $\mathbf{A}^1$-Structure

In algebraic geometry the affine line $\mathbf{A}^1$ plays the rôle of the unit interval. We want the unit interval to be contractible, and again we let the fibrant objects be the "good ones".

Definition 1.1.1 An $\mathbf{A}^1$-*fibrant* object is a locally fibrant object $X \in \mathcal{M}_S$ such that

$$X(\mathbf{A}^1 \times_S -) \to X$$

is a pointwise weak equivalence.

An $\mathbf{A}^1$-*equivalence* is a map $X \to Y \in \mathcal{M}_S$ such that for all $\mathbf{A}^1$-fibrant Z the map

$$\underline{\mathcal{M}}_S(KY, Z) \to \underline{\mathcal{M}}_S(KX, Z)$$

is a pointwise weak equivalence (where K is the cofibrant replacement functor in the pointwise structure).

A map is an $\mathbf{A}^1$-*cofibration* if it is a local cofibration (i.e., a projective cofibration) and an $\mathbf{A}^1$-*fibration* if it has the right lifting property with respect to the trivial $\mathbf{A}^1$-cofibrations.

This is called the $\mathbf{A}^1$-*structure* on $\mathcal{M}_S$ and

Theorem 1.1.2 *The $\mathbf{A}^1$-structure on $\mathcal{M}_S$ is a model category whose homotopy category is equivalent to the "unstable homotopy category" of Voevodsky's lecture.*

We will refer to $\mathcal{M}_S$ with the $\mathbf{A}^1$-structure as "the category of spaces" where "pointed" and "motivic" may be inserted if clarity demands it.

2 Motivic Functors

2.1 Two Questions

1. How can one study (co)homology theories in $\mathcal{M}_S$ (with multiplicative structures) such as motivic cohomology?
2. More generally: How does one study $\mathcal{M}_S$-valued "continuous" and $\mathbf{A}^1$-homotopy invariant functors?

Remark 1. 1. The relevant multiplicative structure comes from the (pointwise) smash. Motivic homology theories must be stable under smashing with the Tate object

$$T = S^1 \wedge \mathbf{G}_m .$$

where $\mathbf{G}_m$ is $A^1 - \{0\}$ pointed at 1.

2. Homology theories should commute with filtered colimits. We only define them on the category $f\mathcal{M}_S \subseteq \mathcal{M}_S$ of finite spaces.

A common answer to the above questions may be found by studying the category

$$\mathcal{F}_S$$

of "continuous" functors $X \colon f\mathcal{M}_S \to \mathcal{M}_S$. Continuous means that X induces a map of morphism spaces $\underline{\mathcal{M}}_S(v, w) \to \underline{\mathcal{M}}_S(X(v), X(w)) \in \mathcal{M}_S$ (not just on the underlying sets). Such functors are referred to by the category theorists as "enriched functors". You have seen before how this is useful when we discussed the Eilenberg-Mac Lane spectrum, and got a map $X \wedge \tilde{\mathbf{Z}}[Y] \to \tilde{\mathbf{Z}}[X \wedge Y]$. This is done by noting that $\tilde{\mathbf{Z}}$ – when considered as a functor on $\mathcal{S}_*$ – induces maps on function spaces, so that you get a chain

$$\mathcal{S}_*(X \wedge Y, X \wedge Y) \cong \mathcal{S}_*(X, \underline{\mathcal{S}}_*(Y, X \wedge Y))$$
$$\to \mathcal{S}_*(X, \underline{\mathcal{S}}_*(\tilde{\mathbf{Z}}[Y], \tilde{\mathbf{Z}}[X \wedge Y])) \cong \mathcal{S}_*(X \wedge \tilde{\mathbf{Z}}[Y], \tilde{\mathbf{Z}}[X \wedge Y]) \,.$$

So if you start with the identity $X \wedge Y = X \wedge Y$ you end up with the desired map. This is the underlying idea for the connection between continuous functors X and spectra: you have maps $S^1 \wedge X(S^n) \to X(S^{n+1})$, and so $n \mapsto X(S^n)$ defines a spectrum.

2.2 Algebraic Structure

A successful study of homology theories and homotopy functors must take care of multiplicative properties. This is provided to us by

Theorem 2.2.1 *(Day) The category $\mathcal{F}_S$ has an internal morphism object and an adjoint associative, commutative and unital smash. The smash of X and Y is built as a filler (up to natural transformation) in*

$$
\begin{array}{ccc}
f\mathcal{M}_S \times f\mathcal{M}_S & \xrightarrow{(u,v) \mapsto X(u) \wedge Y(v)} & \mathcal{M}_S \,. \\
{\scriptstyle (u,v) \mapsto u \wedge v} \downarrow & \nearrow {\scriptstyle v \mapsto (X \wedge Y)(v)} & \\
f\mathcal{M}_S & &
\end{array}
$$

Day would say that $\mathcal{F}_S$ is a "closed symmetric monoidal category" (and he'd be quite right: look it up in your Mac Lane, where you can check up on Kan extensions while you're at it).

Although this definition is rather abstract, in the situations you actually need the smash product it is easy to interpret: the set of maps $X \wedge Y \to Z$ is in one-to-one correspondence with the set of maps $X(u) \wedge Y(v) \to Z(u \wedge v)$ that are natural in u, v. Compare this with the interpretation of the tensor product in terms of bilinear maps.

This is useful since algebras in $\mathcal{F}_S$ are modelled with $\wedge$ just as rings are modelled in $\mathcal{A}b$ by means of the tensor product. The motivic functor corresponding to the integers is *the motivic sphere spectrum* This is simply the inclusion

$$\mathbf{S} \colon f\mathcal{M}_S \to \mathcal{M}_S \, .$$

The multiplication comes in form of an isomorphism $\mathbf{S} \wedge \mathbf{S} \cong \mathbf{S}$. This isomorphism is part of a natural isomorphism $\mathbf{S} \wedge X \cong X$ for any $X \in \mathcal{F}_S$ coming automatically from Day's setup since for any $u, v \in f\mathcal{M}_S$ the continuity of X provides us with a map $u \wedge X(v) \to X(u \wedge v)$.

Just as for the ring of integers is initial among rings, the motivic sphere spectrum $\mathbf{S}$ is the initial algebra in $\mathcal{F}_S$:

Definition 2.2.2 An $\mathbf{S}$-algebra (in $\mathcal{F}_S$) is a continuous functor $A \in \mathcal{F}_S$ together with maps $\mathbf{S} \to A$ (unit) and $A \wedge A \to A$ (multiplication) satisfying the usual associativity and unitality conditions.

With this definition in place, the reader can guess what the definition of A-modules should be, opening up for a algebraic study of the structure of motivic functors.

There are many important $\mathbf{S}$-algebras beside $\mathbf{S}$ itself. The three most prominent are algebraic K-theory, algebraic cobordism theory and the motivic Eilenberg-Mac Lane spectrum. As an example we provide some details on the latter.

2.3 The Motivic Eilenberg-Mac Lane Spectrum

Recall the category Cor_S of correspondences. Its objects are the same as Sm/S, but

$$Cor_S(U, V) =$$
$\mathbf{Z}\{$closed irreducible $C \subseteq U \times V$ finite and surjective over a component of $U\}$.

Let Tr_S be the category of (linear) presheaves $Cor_S^{\mathrm{op}} \to \mathcal{A} = s\mathcal{A}b$. The graph induces a functor

$$Sm/S \to Cor_S \, ,$$

and also a functor $u \colon Tr_S \to \mathcal{M}_S$. This last functor has a left adjoint

$$\gamma \colon \mathcal{M}_S \to Tr_S$$

and γ sends the smash in $\mathcal{M}_S$ to a slightly more complicated tensor which resides in Tr_S.

We define

$$M\mathbf{Z} \colon f\mathcal{M}_S \to \mathcal{M}_S$$

to be the composite

$$f\mathcal{M}_S \subseteq \mathcal{M}_S \xrightarrow{\ \gamma\ } Tr_S \xrightarrow{\ u\ } \mathcal{M}_S \ .$$

The unit map $\mathbf{S} \to M\mathbf{Z}$ is induced by the unit of adjunction $1 \to u\gamma$, and the multiplication $M\mathbf{Z} \wedge M\mathbf{Z} \to M\mathbf{Z}$ comes from the maps $u\gamma(v \wedge w) \cong u(\gamma v \otimes \gamma w) \to u\gamma(v)u\gamma(w)$ which the reader may think of as beefed-up versions of the corresponding maps for the free-forgetful pair between abelian groups and sets.

2.4 Wanted

Model structures on $\mathcal{F}_S = \{$continuous functors $f\mathcal{M}_S \to \mathcal{M}_S\}$ that lift to structures on the categories of algebras and their modules.

It should be noted that such a structure is at present needed just in order to state that there are corresponding algebraic objects on the homotopy categories.

Really, we want one structure to account for homotopy functors among motivic spaces, and one to model the stable situation.

Fact: such structures exist, and the stable situation has a homotopy category which is equivalent to the "stable category" in Voevodsky's talk. This is the second theme of the last section.

3 Model Structures of Motivic Functors and Relation to Spectra

In the last section we defined the category of motivic spaces as the category of simplicial presheaves $Sm/S^{\mathrm{op}} \to \mathcal{S}_*$ together with a suitable model structure for which the fibrant objects

1. are Nisnevich sheaves up to homotopy, and
2. do not see the difference between $\mathbf{A}^1$ and a point.

In this section we will study two structures on the category $\mathcal{F}_S$ of continuous functors $f\mathcal{M}_S \to \mathcal{M}_S$ from finite spaces to spaces. The first one will have as fibrant objects the functors that are blind to the difference between $\mathbf{A}^1$-equivalent spaces, and in the second one they are the functors that think smashing with the Tate object 1.1 should be an invertible operation. Details can be found in [3].

3.1 The Homotopy Functor Model Structure

In practice, one is interested in homotopy functors, i.e., motivic functors that preserve $\mathbf{A}^1$-equivalences. However, the category of homotopy functors is very badly behaved. The trick is to study all motivic functors, but with a model category focusing on the homotopy functors.

Similar to the case for other functor categories (like the projective structure for motivic spaces), there is a model structure – called the *pointwise structure* – on $\mathcal{F}_S$ in which a map $X \to Y$ is a weak equivalence (resp. fibration) if for every $v \in f\mathcal{M}_S$ the map $X(v) \to Y(v) \in \mathcal{M}_S$ is an $\mathbf{A}^1$-equivalence (resp. fibration), and where the cofibrations are the maps having the usual left lifting property.

Let $* \rightarrowtail KX \twoheadrightarrow X$ be the functorial factorization of $* \to X$ in the pointwise structure.

Definition 3.1.1 An *ht-fibrant* object is a pointwise fibrant object $X \in \mathcal{F}_S$ such that for all $\mathbf{A}^1$-equivalences

$$\phi: v \xrightarrow{\sim} w \in f\mathcal{M}_S$$

the induced map

$$X(v) \to X(w) \in \mathcal{M}_S$$

is an $\mathbf{A}^1$-equivalence.

An *ht-equivalence* is a map $X \to Y \in \mathbf{F}_S$ such that for all ht-fibrant $Z \in \mathbf{F}_S$ the induced map

$$\underline{\mathcal{F}}_S(KY, Z) \to \underline{\mathcal{F}}_S(KX, Z)$$

is a pointwise $\mathbf{A}^1$-equivalence.

Being an *ht-cofibration* is the same as being a cofibration in the pointwise structure, and an *ht-fibration* is defined by the right lifting property.

Theorem 3.1.2 *The ht-structure is a model category structure on $\mathcal{F}_S$. This model category lifts to model structures on algebras and modules in $\mathcal{F}_S$.*

Note 3.1.3 *In this structure a map between homotopy functors is an ht-equivalence if and only if it is a pointwise equivalence, and so the homotopy category of the ht-structure is the same as that of the pointwise when restricting to the homotopy functors. Furthermore, any motivic functor is ht-equivalent to a homotopy functor, and the ht-structure gives a functorial replacement making a motivic functor into a homotopy functor.*

We may alternatively characterize the ht-fibrations as the pointwise fibrations $X \to Y$ such that for all $\mathbf{A}^1$-equivalences $v \xrightarrow{\sim} w \in f\mathcal{M}_S$ the square

$$\begin{array}{ccc} X(v) & \longrightarrow & Y(v) \\ \downarrow & & \downarrow \\ X(w) & \longrightarrow & Y(w) \end{array}$$

is a homotopy pullback.

3.2 Motivic Spectra

Recall the Tate object $T = S^1 \wedge \mathbf{G}_m$ 1.1. Motivic cohomology, algebraic K-theory and so on are stable with respect to both S^1 and $\mathbf{G}_m$, and are represented by T-spectra. Naïvely one can model T-spectra in the same way as we did for spectra in spaces in III.5.

Definition 3.2.1 A T-spectrum is a sequence

$$E^0, E^1, \cdots \in \mathcal{M}_S = \mathcal{S}_*^{Sm/S^{\mathrm{op}}}$$

together with structure maps

$$T \wedge E^n \to E^{n+1} .$$

T-spectra are also referred to as *motivic spectra.*

As in the simplicial set case, these form a category $\mathcal{S}pt_S$ that comes equipped with a model structure whose homotopy category is equivalent to the "motivic stable homotopy category" of Voevodsky's talk [22]. The nicest way (following Hovey [10]) to describe the stable equivalences is the following: construct a "stably fibrant replacement functor", along the lines of the Q in 5.2 (makes the spectrum into an "Ω-spectrum"), and declare that a map $X \to Y$ of spectra is a *stable equivalence* if $(QX)^n \to (QY)^n$ is an $\mathbf{A}^1$-equivalence for each n. Similarly, stable fibrations are defined, giving rise to the *stable structure*.

In search for a deeper structure on this category we may mimic one of the ways of of constructing such structures for ordinary spaces. Among approaches we may mention symmetric spectra [11], S-modules [5] and simplicial functors [15]. Since I personally like the last approach best, this is what we have prepared for through our focus on motivic functors. I give a brief outline how this goes leaving details to [3].

3.3 The Connection $\mathcal{F}_S \to \mathcal{S}pt_S$

Let $T^k = T \wedge T^{k-1}$ and $T^0 = S^0$ (the constant two-point functor: it is the unit of the $\wedge$-structure on $\mathcal{M}_S$). For any continuous $X \colon f\mathcal{M}_S \to \mathcal{M}_S$ let the *evaluation* $evX \in \mathcal{S}pt_S$ be the spectrum with n-th term $X(T^n)$. The beforementioned map $T \wedge X(T^n) \to X(T^{n+1})$ coming from the continuity of X ensures that evX actually is a spectrum. Obviously $ev \colon \mathcal{F}_S \to \mathcal{S}pt_S$ preserves both limits and colimits.

The idea is to transport the stable equivalences from $\mathcal{S}pt_S$ to $\mathcal{F}_S$, and this almost forces the stable structure on $\mathcal{F}_S$. More precisely, to check whether $X \to Y$ is a *stable equivalence* one first uses the ht-structure to replace X and Y by homotopy functors, and then asks whether the induced map of T-spectra is a stable equivalence.

Call the left adjoint of ev

$$F \colon \mathcal{S}pt_S \to \mathcal{F}_S .$$

Theorem 3.3.1 *There is a model category structure, called the stable structure, on $\mathcal{F}_S$ such that*

$$F: \mathcal{S}pt_S \rightarrow \mathcal{F}_S$$

is a Quillen equivalence.

Furthermore, the smash in $\mathcal{F}_S$ gives a smash in the stable homotopy category, agreeing with that discussed in the other talks. The stable model structure induces model category structures on algebras and modules over $\mathcal{F}_S$.

In algebraic topology there is an interesting interpretation of the stable structure in terms of linear functors in the sense of Goodwillie calculus. This connection is not as straight-forward in the motivic world. In [3] what we here have called the stable structure is called the spherewise structure.

References

1. A. K. Bousfield and E. M. Friedlander. Homotopy theory of Γ-spaces, spectra, and bisimplicial sets. In *Geometric applications of homotopy theory (Proc. Conf., Evanston, Ill., 1977), II*, Vol. 658 of *Lecture Notes in Math.*, pp. 80–130. Springer, Berlin, 1978.
2. A. K. Bousfield and D. M. Kan. *Homotopy limits, completions and localizations. Lecture Notes in Mathematics*, Vol. 304. Springer-Verlag, Berlin, 1972.
3. B. I. Dundas, O. Röndigs, and P. A. Østvær. Motivic functors. *Doc. Math.*, 8: 489–525 (electronic), 2003.
4. W. G. Dwyer, P. S. Hirschhorn, D. M. Kan, and J. H. Smith. *Homotopy limit functors on model categories and homotopical categories*, Vol. 113 of *Mathematical Surveys and Monographs*. American Mathematical Society, Providence, RI, 2004.
5. A. D. Elmendorf, I. Kriz, M. A. Mandell, and J. P. May. *Rings, modules, and algebras in stable homotopy theory*, Vol. 47 of *Mathematical Surveys and Monographs*. American Mathematical Society, Providence, RI, 1997. With an appendix by M. Cole.
6. P. G. Goerss and J. F. Jardine. *Simplicial homotopy theory*, Vol. 174 of *Progress in Mathematics*. Birkhäuser Verlag, Basel, 1999.
7. A. Hatcher. *Algebraic topology*. Cambridge University Press, Cambridge, 2002.
8. P. S. Hirschhorn. *Model categories and their localizations*, Vol. 99 of *Mathematical Surveys and Monographs*. American Mathematical Society, Providence, RI, 2003.
9. M. Hovey. *Model categories*, Vol. 63 of *Mathematical Surveys and Monographs*. American Mathematical Society, Providence, RI, 1999.
10. M. Hovey. Spectra and symmetric spectra in general model categories. *J. Pure Appl. Algebra*, 165(1): 63–127, 2001.
11. M. Hovey, B. Shipley, and J. Smith. Symmetric spectra. *J. Amer. Math. Soc.*, 13(1): 149–208, 2000.
12. J. F. Jardine. Motivic symmetric spectra. *Doc. Math.*, 5: 445–553 (electronic), 2000.
13. M. Levine. Lectures in nordfjordeid. In *Summer school on motivic homotopy theory*. This volume.

14. E. L. Lima. The Spanier-Whitehead duality in new homotopy categories. *Summa Brasil. Math.*, 4: 91–148, 1959.
15. M. Lydakis. Simplicial functors and stable homotopy theory. Preprint 98-049, SFB 343, Bielefeld, June 1998.
16. S. M. Lane. *Categories for the working mathematician*, Vol. 5 of *Graduate Texts in Mathematics*. Springer-Verlag, New York, second edition, 1998.
17. J. Peter May. *Simplicial objects in algebraic topology*. Chicago *Lectures in Mathematics*. University of Chicago Press, Chicago, IL, 1992. Reprint of the 1967 original.
18. J. R. Munkres. *Topology: a first course*. Prentice-Hall Inc., Englewood Cliffs, N.J., 1975.
19. D. Quillen. Rational homotopy theory. *Ann. of Math. (2)*, 90: 205–295, 1969.
20. D. Quillen. On the (co-)homology of commutative rings. In *Applications of Categorical Algebra (Proc. Sympos. Pure Math., Vol. XVII, New York, 1968)*, pp. 65–87. Amer. Math. Soc., Providence, R.I., 1970.
21. D. G. Quillen. *Homotopical algebra. Lecture Notes in Mathematics*, No. 43. Springer-Verlag, Berlin, 1967.
22. V. Voevodsky, P. A. Østvær, and O. Röndigs. Voevodsky's lectures in nordfjordeid. In *Summer school on motivic homotopy theory*. This volume.

Index

Background from Algebraic Geometry

Marc Levine*

Department of Mathematics, Northeastern University, Boston, MA 02115, USA
marc@neu.edu

* The author gratefully acknowledges the support of the Humboldt Foundation through the Wolfgang Paul Program, and support of the NSF via grants DMS-0140445 and DMS-0457195.

Elementary Algebraic Geometry

In this first part, we give a quick overview of some of the foundational material of elementary algebraic geometry needed for a study of motivic homotopy theory. All of this material is well-known and excellently discussed in numerous texts; our goal is to collect the main facts to give the reader a convenient first introduction and quick reference. For this reason, many of the proofs will be only sketched or completely omitted. For further details, we suggest the reader take a look at [4], [10], [14] for the commutative algebra and [7], [9] or [12] for the algebraic geometry; for a more analytic point of view, we suggest [5].

In part 2, we discuss the notions of sheaves and presheaves in the setting of a Grothendieck topology, as needed for the construction and understanding of motivic homotopy theory.

This text is a selection of material I presented in the summer school on Motivic Homotopy Theory at Nordfjordeid, intended as a quick introduction to the algebraic geometry and sheaf-theory needed for constructions in motivic homotopy theory. I would like to thank Bjørn Jahren for all his work organizing the summer school and this book, Bjørn Dundas and Vladimir Voevodsky for their excellent lectures, and the participants for creating an enthusiastic and stimulating atmosphere.

1 The Spectrum of a Commutative Ring

We begin our introduction by discussing the geometric viewpoint of commutative algebra. All rings will be assumed to be commutative and with unit.

1.1 Ideals and Spec

Let A be a ring. Recall that an ideal $\mathfrak{p} \subset A$ is a *prime* ideal if $\mathfrak{p}$ is a proper ideal (i.e., $\mathfrak{p} \neq A$) and

$$ab \in \mathfrak{p}, a \notin \mathfrak{p} \implies b \in \mathfrak{p}.$$

This property easily extends from elements to ideals: If I and J are ideals of A, we let IJ be the ideal generated by products ab with $a \in I$, $b \in J$. Then, if $\mathfrak{p}$ is a prime ideal, we have

$$IJ \subset \mathfrak{p}, I \not\subset \mathfrak{p} \Longrightarrow J \subset \mathfrak{p}.$$

Since $IJ \subset I \cap J$, we have as well

$$I \cap J \subset \mathfrak{p}, I \not\subset \mathfrak{p} \Longrightarrow J \subset \mathfrak{p}.$$

Example 1.1 *A ring A is an* integral domain *(or just a domain) if for all $a, b \in A$,*

$$ab = 0, \ a \neq 0 \Longrightarrow b = 0.$$

In terms of ideals, A is an integral domain if and only if the zero-ideal (0) is a prime ideal, and for an arbitrary ring A, an ideal I of A is prime if and only if A/I is an integral domain.

In addition to product and intersection, we have the operation of sum: if $\{I_\alpha \mid \alpha \in \mathcal{A}\}$ is a set of ideals of A, we let $\sum_\alpha I_\alpha$ be the smallest ideal of A containing all the ideals I_α. One easily sees that

$$\sum_\alpha I_\alpha = \{\sum_\alpha x_\alpha \mid x_\alpha \in I_\alpha \text{ and almost all } x_\alpha = 0\}.$$

We let $\mathrm{Spec}\,(A)$ denote the set of prime ideals of A:

$$\mathrm{Spec}\,(A) := \{\mathfrak{p} \subset A \mid \mathfrak{p} \text{ is a prime ideal}\}.$$

For a subset S of A, we set

$$V(S) = \{\mathfrak{p} \in \mathrm{Spec}\,(A) \mid \mathfrak{p} \supset S\}.$$

We note the following properties of the operation V:

1. Let S be a subset of A, and let $(S) \subset A$ be the ideal generated by S. Then $V(S) = V((S))$.
2. Let $\{I_\alpha \mid \alpha \in \mathcal{A}\}$ be a set of ideals of A. Then $V(\sum_\alpha I_\alpha) = \cap_\alpha V(I_\alpha)$.
3. Let $I_1, \ldots, I_N$ be ideals of A. Then $V(\cap_{j=1}^N I_j) = \cup_{j=1}^N V(I_j)$.
4. $V(0) = \mathrm{Spec}\,(A)$, $V(A) = \emptyset$.
5. Let $I \subset A$ be an ideal, and let $\pi : A \to A/I$ be the quotient map. Then sending $\bar{J} \subset A/I$ to $J := \pi^{-1}(\bar{J})$ gives a bijection $\pi^* : \mathrm{Spec}\, A/I \to V(I) \subset \mathrm{Spec}\, A$.

$$(1)$$

1.2 The Zariski Topology

Definition 1.2 *The* Zariski topology *on* $\operatorname{Spec}(A)$ *is the topology for which the closed subsets are exactly the subsets of the form* $V(I)$, *I an ideal of A.*

It follows from the properties (1) that this really does define a topology on the set $\operatorname{Spec}(A)$.

The Zariski topology is quite different from other familiar topologies, like the metric topology on $\mathbb{R}^n$. For instance, the points of $\operatorname{Spec} A$ are the prime ideals of A, which can be ordered by inclusion. Only the points of $\operatorname{Spec} A$ corresponding to maximal ideals of A are closed, since the closure of the point $[\mathfrak{p}] \in \operatorname{Spec} A$ corresponding to a prime idea $\mathfrak{p} \subset A$ is $V(\mathfrak{p})$. In particular the Zariski topology is not Hausdorff if there is a non-trivial containment of prime ideals $\mathfrak{p} \subset \mathfrak{q}$. For example, if A is an integral domain, so that the zero-ideal (0) is prime, then $\operatorname{Spec} A = V((0))$, so the closure of the point $[(0)]$ is all of $\operatorname{Spec} A$. Thus for a domain A, $\operatorname{Spec} A$ is not Hausdorff unless A is a field.

On the positive side, the spaces $\operatorname{Spec} A$ often behave much like compact topological spaces, even though they usually have no hope of being compact (since they usually aren't even Hausdorff).

Definition 1.3 *(1) A topological space $\mathcal{T}$ is* noetherian *if every sequence of closed subspaces*
$$F_0 \supset F_1 \supset F_2 \supset \dots$$
is eventually constant.

(2) A commutative ring A is noetherian *if every sequence of ideals*
$$I_0 \subset I_1 \subset I_2 \subset \dots$$
is eventually constant.

So, $\operatorname{Spec} A$ is a noetherian space if A is a noetherian ring (although not conversely, for example the ring $k[x_1, x_2, \dots]/(x_1^2, x_2^2, \dots)$). As we will see later, a quotient of a polynomial algebra over a field
$$A = F[x_1, \dots, x_n]/I$$
is noetherian, so $\operatorname{Spec} F[x_1, \dots, x_n]/I$ is also noetherian.

If $\mathcal{T}$ is a noetherian space and $\mathcal{T} = \cup_{i=0}^{\infty} U_i$ for a family of open subsets $U_0, U_1, \dots$, then the fact that the sequence of closed subsets
$$\mathcal{T} \setminus U_0 \supset \mathcal{T} \setminus U_0 \cup U_1 \supset \dots$$
is eventually constant means $\mathcal{T} = \cup_{i=0}^{N} U_i$ for some N. One calls a topological space with this property *quasi-compact*. So, $\operatorname{Spec} A$ is quasi-compact if A is noetherian.

Also, if $\mathcal{T}$ is a noetherian topological space, each closed subset C is a finite union of *irreducible* closed subsets: $C = \cup_{i=1}^r C_i$, where C is irreducible if it can't be written as a union of two proper closed subsets. This fact follows easily from the noetherian property. For $\mathcal{T} = \operatorname{Spec} A$, the irreducible closed subsets are exactly the closures of single points, i.e., subsets of the form $V(\mathfrak{p})$ with $\mathfrak{p}$ a prime ideal. Thus, for an ideal $I \subset A$,

$$V(I) = \cup_{i=1}^n V(\mathfrak{p}_i)$$

for finitely many prime ideals $\mathfrak{p}_1, \ldots, \mathfrak{p}_n$. In particular, this says that each ideal I in a noetherian ring A has finitely many *minimal* prime ideals containing it (minimal in the sense of containment).

A similar type of argument gives the well-known fact that a commutative ring A is noetherian if and only if all ideals are finitely generated. See section 4.1 for more details on noetherian rings.

Examples 1.4 *(1) Let F be a field. Then (0) is the unique proper ideal in F, and is prime since F is an integral domain. Thus $\operatorname{Spec}(F)$ is the one-point space $\{[(0)]\}$ (we use $[-]$ to distinguish a prime ideal from a point in Spec).*

(2) Again, let F be a field, and let $A = F[t]$ be a polynomial ring in one variable t. A is an integral domain, so (0) is a prime ideal. A is a unique factorization domain (UFD), which means that the ideal (f) generated by an irreducible polynomial f is a prime ideal. In fact, each non-zero proper prime ideal in A is of the form (f), $f \neq 0$, f irreducible. If we take f monic, then different f's give different ideals, so we have

$$\operatorname{Spec}(A) = \{[(0)]\} \cup \{[(f)] \mid f \in A \text{ a monic irreducible polynomial}\}.$$

Clearly $(0) \subset (f)$ for all such f; there are no other containment relations as the ideal (f) is maximal (i.e., not contained in any other proper ideal) if f is irreducible, $f \neq 0$. Thus the closure of $[(0)]$ is all of $\operatorname{Spec}(A)$, and the other points $[(f)] \in \operatorname{Spec}(A)$, f irreducible and monic, are closed points.

(3) The affine line over a field. Suppose F is algebraically closed, e.g., $F = \mathbb{C}$. Then an irreducible monic polynomial $f \in F[t]$ is necessarily linear, hence $f = t - a$ for some $a \in F$. We thus have a 1-1 correspondence between the closed points of $\operatorname{Spec}(F[t])$ and the set F. For this reason, $\operatorname{Spec}(F[t])$ is called the affine line over F (even if F is not algebraically closed), written $\mathbb{A}_F^1$.

(4) $\operatorname{Spec} \mathbb{Z}$. This is the first "arithmetic" example. $\operatorname{Spec} \mathbb{Z}$ has the generic point $[(0)]$ and closed points $[(p)]$ for each prime number p.

1.3 Functorial Properties

The operation $A \mapsto \operatorname{Spec}(A)$ actually defines a contravariant functor from the category of commutative rings to topological spaces. In fact, let $\phi : A \to B$

be a homomorphism of commutative rings, and let $\mathfrak{p} \subset B$ be a proper prime ideal. Then it follows directly from the definition that $\phi^{-1}(\mathfrak{p})$ is a prime ideal of A, and is proper, since $1_A \in \phi^{-1}(\mathfrak{p})$ implies $1_B = \phi(1_A)$ is in $\mathfrak{p}$, which implies $\mathfrak{p} = B$. Thus, sending $\mathfrak{p}$ to $\phi^{-1}(\mathfrak{p})$ defines the map of sets

$$\hat{\phi} : \operatorname{Spec}(B) \to \operatorname{Spec}(A).$$

In addition, if I is an ideal of A, then $\phi^{-1}(\mathfrak{p}) \supset I$ for some $\mathfrak{p} \in \operatorname{Spec}(B)$, if and only if $\mathfrak{p} \supset \phi(I)$. Thus $\hat{\phi}^{-1}(V(I)) = V(\phi(I))$, hence $\hat{\phi}$ is continuous.

The space $\operatorname{Spec}(A)$ encodes lots (but not all) of the information regarding the ideals of A. For example, let $I \subset A$ be an ideal. We have the quotient ring A/I and the canonical surjective ring homomorphism $\phi_I : A \to A/I$. If $\bar{J} \subset A/I$ is an ideal, we have the inverse image ideal $J := \phi^{-1}(\bar{J})$; sending $\bar{J}$ to J is then a bijection between the ideals of A/I and the ideals J of A with $J \supset I$. We have the following extension of the property (1)5:

Lemma 1.5 *The map $\hat{\phi}_I : \operatorname{Spec}(A/I) \to \operatorname{Spec}(A)$ gives a homeomorphism of $\operatorname{Spec}(A/I)$ with the closed subspace $V(I)$ of $\operatorname{Spec}(A)$.*

On the other hand, one can have $V(I) = V(J)$ even if $I \neq J$. For an ideal I, the *radical* of I is the ideal

$$\sqrt{I} := \{x \in A \mid x^n \in I \text{ for some integer } n \geq 1\}.$$

It is not hard to see that $\sqrt{I}$ really is an ideal. Clearly, if $\mathfrak{p}$ is prime, and $x^n \in \mathfrak{p}$ for some $n \geq 1$, then $x \in \mathfrak{p}$, so $V(I) = V(\sqrt{I})$, but it is easy to construct examples of ideals I with $I \neq \sqrt{I}$ (e.g. $I = (t^2) \subset F[t]$, F a field). In terms of Spec, the quotient map $A/I \to A/\sqrt{I}$ induces the homeomorphism

$$\operatorname{Spec}(A/\sqrt{I}) \to \operatorname{Spec}(A/I).$$

In fact, $\sqrt{I} \supset I$ is the largest ideal with this property, since we have the formula

$$\sqrt{I} = \cap_{\mathfrak{p} \supset I} \mathfrak{p}. \tag{2}$$

1.4 Naive Algebraic Geometry and Hilbert's Nullstellensatz

As the definition of $\operatorname{Spec} A$ is very general, it's a good idea to see what happens in the "geometric" case". Take an algebraically closed field k, and take some polynomials

$$f_1, \ldots, f_r \in k[x_1, \ldots, x_n].$$

The geometric data this determines is the set of solutions to the equations $f_i = 0$, so let

$$V_{geom}(f_1, \ldots, f_r) := \{a = (a_1, \ldots, a_n) \in k^n \mid f_1(a) = \ldots = f_r(a) = 0\}$$

We also have the ideal $(f_1, \ldots, f_r)$ and the closed subset $V((f_1, \ldots, f_r))$ of $\operatorname{Spec} k[x_1, \ldots, x_n]$. Hilbert's Nullstellensatz relates these two.

Theorem 1.6 (Hilbert's Nullstellensatz, first form) *For k an algebraically closed field, the maximal ideals of $k[x_1, \ldots, x_n]$ are in one-to-one correspondence with k^n, by*

$$(a_1, \ldots, a_n) \mapsto (x_1 - a_1, \ldots, x_n - a_n).s$$

To use this to relate $V_{geom}(f_1, \ldots, f_r)$ and $V((f_1, \ldots, f_r))$, we just have to note

Lemma 1.7 *Take $a = (a_1, \ldots, a_n) \in k^n$. We have a containment of ideals*

$$(f_1, \ldots, f_r) \subset (x_1 - a_1, \ldots, x_n - a_n)$$

if and only if $f_1(a) = \ldots, f_r(a) = 0$.

In fact, f_i is in $(x_1 - a_1, \ldots, x_n - a_n)$ if and only if we can write

$$f_i = \sum_j g_{ij}(x_j - a_j), \ i = 1, \ldots, r,$$

for some $g_{ij} \in k[x_1, \ldots, x_r]$, so one implication is clear. The other follows by dividing by $(x_1 - a_1)$, dividing the remainder by $(x_2 - a_2)$, etc., yielding an identity

$$f_i = \sum_j g_{ij}(x_j - a_j) + c.$$

c is a constant by reason of degree; evaluating at $x = a$, we see that $c = 0$.

The upshot is that $(a_1, \ldots, a_n)$ is in $V_{geom}(f_1, \ldots, f_r)$ if and only if $(x_1 - a_1, \ldots, x_n - a_n)$ is in $V((f_1, \ldots, f_r))$. If we let $\mathrm{Spec}_{max} A \subset \mathrm{Spec}\, A$ denote the subset of maximal ideals of A, then Hilbert's Nullstellensatz gives a bijection

$$k^n \cong \mathrm{Spec}_{max} k[x_1, \ldots, x_n]$$

and under this identification, we have

$$V_{geom}(f_1, \ldots, f_r) = V((f_1, \ldots, f_r)) \cap \mathrm{Spec}_{max} k[x_1, \ldots, x_n].$$

This raises the question: why not just use $\mathrm{Spec}_{max} A$ instead of the more complicated $\mathrm{Spec}\, A$? There are lots of reasons, but one obvious one is that, as we've seen, $A \mapsto \mathrm{Spec}\, A$ is easily made into a functor, but there is no reasonable way that this can be done with Spec_{max}.

Example 1.8 Affine spaces. *The affine space of dimension n over a field F is $\mathrm{Spec}\, F[x_1, \ldots, x_n]$, written $\mathbb{A}_F^n$. We've just seen that*

$$\mathrm{Spec}_{max} F[x_1, \ldots, x_n] \subset \mathbb{A}_F^n$$

is F^n, for F algebraically closed, which justifies the terminology.

1.5 Krull Dimension, Height One Primes and the UFD Property

To give a little more feeling about Spec, we recall some elementary facts from commutative algebra.

Let A be a commutative ring. If we order the prime ideals of A by inclusion, we get a measure of how big a prime ideal $\mathfrak{p}$ is: $\mathfrak{p}$ has *height* m if there is a proper chain of prime ideals

$$\mathfrak{p}_0 \subset \mathfrak{p}_1 \subset \ldots \subset \mathfrak{p}_m = \mathfrak{p}$$

and no longer chain. For $A = F[x_1, \ldots, x_n]/I$, it turns out that all maximal such chains have the same length (although there are bizarre rings for which this is not the case), and the maximal ideals of $F[x_1, \ldots, x_n]$ all have height n. Thus, for many rings, a reasonable measure of their dimension is the length of a maximal chain of prime ideals.

Definition 1.9 *(1) Let A be a noetherian commutative ring. The* Krull dimension *of A is the maximum of the set of heights of prime ideals of A (or infinity if there is no maximum or if some prime ideal has infinite height).*

(2) Let T be a noetherian topological space. The dimension *of T is the maximum of the length n of a chain of irreducible closed subsets*

$$T_0 \supset T_1 \supset \ldots \supset T_n.$$

In particular, $\dim A = \dim \operatorname{Spec} A$.

Recall that a *unit* in A is an element u with a multiplicative inverse: $uv = 1$. An element f in A is *irreducible* if f is not a unit and $f = gh$ implies that either g or h is a unit. If A is a domain, an element $f \in A \setminus \{0\}$ is called *prime* if f is not a unit, and if f divides gh, then f divides g or f divides h.

A domain A is a *unique factorization domain* if each element f in A admits a finite factorization into irreducibles, $f = \prod_i f_i$ which is unique in the following sense: if

$$\prod_{i=1}^{n} f_i = \prod_{j=1}^{m} g_j$$

with all the f_i and g_j irreducible, then $n = m$, and after reordering, there are units $u_i \in A$ with $f_i = u_i g_i$, $i = 1, \ldots, n$.

The basic facts about UFD's are

Theorem 1.10 *Let A be a noetherian domain.*

(1) Each element $f \in A \setminus \{0\}$ can be written as a finite product $f = \prod_i f_i$, with the f_i irreducible.

(2) A prime element is always irreducible and the following are equivalent:

(a) each irreducible in A is prime
(b) each height one prime ideal of A is principal: $\mathfrak{p} = (f)$ for some element $f \in A$.
(c) A is a UFD

Example 1.11 *For F a field, a polynomial algebra $F[x_1, \ldots, x_n]$ is a UFD: this follows from Gauss' lemma. In general, a quotient ring $F[x_1, \ldots, x_n]/\mathfrak{p}$ is not a UFD; the simplest example is the "universal" one: $A := F[X, Y, Z, W]/(XY - ZW)$. It is easy to see that X, Y, Z and W are irreducible in A (use the fact that the ideal is homogeneous of degree 2), but clearly $XY = ZW$ in A is a non-unique factorization.*

Examples 1.12 *The affine plane over a field. Considering $\mathbb{A}^2_F$, we find three kinds of points: the generic point $[(0)]$, the height one points $[(f)]$, with $f(x, y)$ an irreducible polynomial, and the closed points $[\mathfrak{m}]$ (which are all of the form $(x - a, y - b)$ if F is algebraically closed).*

For F algebraically closed, it's easy to understand $V(f(x, y))$ (say for f irreducible). One has the prime (f) with closure $V((f))$; by the Nullstellensatz, $V(f)$ consists of the generic point $[(f)]$ and the closed points $[(x - a, y - b)]$ with $f(a, b) = 0$.

So, $\operatorname{Spec} F[x, y]$ consists of not only the "obvious" points (a, b), but one point for each irreducible curve $f = 0$ in the plane and one more point $[(0)]$ "representing" the whole plane. The closure relation tells you which points (a, b) are solutions of the equations $f = 0$.

The affine line over $\mathbb{Z}$. Consider $\mathbb{A}^1_{\mathbb{Z}} := \operatorname{Spec} \mathbb{Z}[x]$. This looks a bit like $\mathbb{A}^2_F$, with some important differences. $\mathbb{Z}[x]$ is also a UFD and has Krull dimension two, so we have the generic point $[(0)]$, the codimension one points $[(f)]$ for irreducible $f \in \mathbb{Z}[x]$ and the codimension two points $[\mathfrak{m}]$ for $\mathfrak{m}$ a maximal ideal. There are two distinct types of codimension one points $[(f)]$: those coming from irreducible f of degree > 0 and those of the form $[(p)]$, $p \in \mathbb{Z}$ prime. The first type don't disappear when you pass from $\mathbb{Z}[x]$ to $\mathbb{Q}[x]$, while $[(p)]$ lives in characteristic $p > 0$. The maximal ideals $\mathfrak{m}$ all have quotient rings $\mathbb{Z}[x]/\mathfrak{m}$ a finite field, so the point $[\mathfrak{m}]$ lives in finite characteristic as well.

Examples 1.13 *Affine spaces over a field. For $\mathbb{A}^n_F$, the situation is similar to $\mathbb{A}^2_F$, but more complicated. We have points $[\mathfrak{p}]$ for each prime ideal $\mathfrak{p}$, with closure $V(\mathfrak{p})$ having dimension somewhere from 0 to n. Choosing generators $f_1, \ldots, f_r$ for a given prime ideal $\mathfrak{p}$, we again have (if F is algebraically closed)*

$$V(\mathfrak{p}) \cap \operatorname{Spec}_{max} = \{a = (a_1, \ldots, a_n) \mid f_1(a) = \ldots, f_r(a) = 0\},$$

or, without choosing generators

$$V(\mathfrak{p}) \cap \operatorname{Spec}_{max} = \{a = (a_1, \ldots, a_n) \mid f(a) = 0 \text{ for all } f \in \mathfrak{p}\},$$

Here we identify $\text{Spec}_{max} F[x_1, \ldots, x_n]$ *with* F^n *by the Nullstellensatz.*

$F[x_1, \ldots, x_n]$ *is a UFD, so the height one primes are all principal; in terms of the irreducible closed subsets of* $\mathbb{A}^n_F$, *this says that each irreducible hypersurface (i.e. closed subset of codimension one) is of the form* $V(f)$ *for* f *an irreducible polynomial in* $F[x_1, \ldots, x_n]$. *However, it is not true the each height* r *prime is generated by* r *elements; the most one can say in general is that it takes at least* r *elements to generate a height* r *prime. This phenomenon occurs at the first place the numerology says it can, namely, for a height two prime in* $F[x, y, z]$, *geometrically, a curve in* $\mathbb{A}^3_F$. *One example of this is the curve parametrized by*

$$t \mapsto (t(t-1)^2(t-2)^2, t^2(t-1)(t-2)^2, t^2(t-1)^2(t-2)).$$

(2) For A *of the form* $F[x_1, \ldots, x_n]/I$, $\text{Spec}\, A$ *is identified with the closed subspace* $V(I)$ *of* $\mathbb{A}^n_F$, *so everything we've seen about* $\mathbb{A}^n_F$ *carries over to* $\text{Spec}\, A$: *we have the irreducible subset* $V(\mathfrak{p})$, $\mathfrak{p} \subset A$ *prime, with* $V(\mathfrak{p})$ *the closure of its "generic point"* $[\mathfrak{p}]$ *and with* $V(\mathfrak{p}) \subset V(\mathfrak{q})$ *exactly when* $\mathfrak{p} \supset \mathfrak{q}$. *For* F *algebraically closed, we recover via Hilbert's Nullstellensatz the naive notion of algebraic geometry as the study of solutions of polynomial equation.*

1.6 Open Subsets and Localization

We've seen that a closed subset $V(I)$ of $\text{Spec}\, A$ is itself the spectrum of a ring, namely, $V(I) = \text{Spec}\, A/I$, but this is not the case in general for open subsets of $\text{Spec}\,(A)$. There is however a natural basis of the topology which does have an algebraic interpretation along these lines. We first consider a more general construction, called localization.

Let S be a subset of A, closed under multiplication and containing 1. Form the *localization* of A with respect to S as the ring of "fractions" a/s with $a \in A$, $s \in S$, where we identify two fraction a/s, a'/s' if there is a third $s'' \in S$ with

$$s''(s'a - sa') = 0.$$

We multiply and add the fractions by the usual rules. (If A is an integral domain, the element s'' is superfluous, but in general it is needed to make sure that the addition of fractions is well-defined). We denote this ring by $S^{-1}A$; sending a to $a/1$ defines the ring homomorphism $\phi_S : A \to S^{-1}A$. If f is an element of A, we may take $S = S(f) := \{f^n \mid n = 0, 1, \ldots\}$, and write A_f for $S(f)^{-1}A$. The homomorphism ϕ_S is universal for ring homomorphisms $\psi : A \to B$ such that B is commutative and $\psi(S)$ consists of units in B.

Note that this operation allows one to invert elements one usually doesn't want to invert, for example 0 or non-zero nilpotent elements. In this extreme case, we end up with the 0-ring; in general the ring homomorphism ϕ_S is not injective. However, all is well if we don't invert *zero-divisors*, i.e., an element $a \neq 0$ such that there is a $b \neq 0$ with $ab = 0$. The most well-know case of localization is of course the formation of the quotient field of an integral domain A, where we take $S = A \setminus \{0\}$.

What does localization do to ideals? If $\mathfrak{p} \neq A$ is prime, there are two possibilities: If $\mathfrak{p} \cap S = \emptyset$, then the image of $\mathfrak{p}$ in $S^{-1}A$ generates a proper prime ideal. In fact, if $\mathfrak{q} \subset S^{-1}A$ is the ideal generated by $\phi_S(\mathfrak{p})$, then

$$\mathfrak{p} = \phi_S^{-1}(\mathfrak{q}).$$

If $\mathfrak{p} \cap S \neq \emptyset$, then clearly the image of $\mathfrak{p}$ in $S^{-1}A$ contains invertible elements, hence generates the unit ideal $S^{-1}A$. Thus (see [4] for details),

Lemma 1.14 *Let $S \subset A$ be a multiplicatively closed subset containing 1. Then $\hat{\phi}_S : \mathrm{Spec}\,(S^{-1}A) \to \mathrm{Spec}\,(A)$ gives a homeomorphism of $\mathrm{Spec}\,(S^{-1}A)$ with the complement of $\cup_{g \in S} V((g))$ in $\mathrm{Spec}\,(A)$.*

In general, the subspace $\hat{\phi}_S(\mathrm{Spec}\,(S^{-1}A))$ is not open, but if $S = S(f)$ for some $f \in A$, then the image of $\hat{\phi}_S$ is the open complement of $V((f))$. An open subset of this form is called *principal*; we write $\mathrm{Spec}\,(A)_f$ for $\mathrm{Spec}\,(A) \setminus V(f)$, A_f for $S(f)^{-1}A$ and $\phi_f : A \to A_f$ for the localization homomorphism.

Lemma 1.15 *The principal open subsets $\mathrm{Spec}\,(A)_f$ form a basis for the Zariski topology on $\mathrm{Spec}\,A$.*

This follows from the obvious identities

$$\mathrm{Spec}\,A \setminus V(I) = \cup_{f \in I} \mathrm{Spec}\,(A)_f$$
$$\mathrm{Spec}\,(A)_f \cap \mathrm{Spec}\,(A)_g = \mathrm{Spec}\,(A)_{fg}.$$

Lemma 1.14 yields as a special case

Proposition 1.16 *Sending $\mathfrak{q} \subset A_f$ to $\phi_f^{-1}(\mathfrak{q}) \subset A$ defines a homeomorphism $\mathrm{Spec}\,(A_f) \to \mathrm{Spec}\,(A)_f$.*

Examples 1.17 *(1) The open subset $\mathrm{Spec}\,F[x, x^{-1}]$ of $\mathbb{A}_F^1 = \mathrm{Spec}\,F[x]$ is the principal open subset defined by x. Note that*

$$F[x, x^{-1}] \cong F[x, y]/(xy - 1)$$

so the open subset $\mathrm{Spec}\,(F[x])_x$ of $\mathbb{A}_F^1$ is also the closed subset $V(xy - 1)$ of $\mathbb{A}_F^2$.

(2) Let $A = F[x_1, \ldots, x_n]/I$ and let $f \in A$ be the image of some element $\tilde{f} \in F[x_1, \ldots, x_n]$. The principal open subset $\mathrm{Spec}\,(A)_f$ is homeomorphic to the closed subset $V((I, \tilde{f}x_{n+1} - 1))$ of $\mathbb{A}_F^{n+1}$, since

$$A[1/f] \cong A[y]/(fy - 1).$$

2 Ringed Spaces

2.1 Presheaves and Sheaves on a Space

Let T be a topological space, and let $\mathrm{Op}(T)$ be the category with objects the open subsets of T and morphisms $V \to U$ corresponding to inclusions $V \subset U$. Recall that a *presheaf* $\mathcal{S}$ (of abelian groups) on T is a functor $\mathcal{S} : \mathrm{Op}(T)^{\mathrm{op}} \to$ **Ab**. For $V \subset U$, we often denote the homomorphism $\mathcal{S}(U) \to \mathcal{S}(V)$ by $\mathrm{res}_{V,U}$. A presheaf $\mathcal{S}$ is a *sheaf* if for each open covering $U = \cup_\alpha U_\alpha$, the sequence

$$0 \to \mathcal{S}(U) \xrightarrow{\Pi_\alpha \, \mathrm{res}_{U_\alpha,U}} \prod_\alpha \mathcal{S}(U_\alpha)$$

$$\xrightarrow{\Pi_{\alpha,\beta} \, \mathrm{res}_{U_\alpha \cap U_\beta, U_\alpha} - \mathrm{res}_{U_\alpha \cap U_\beta, U_\beta}} \prod_{\alpha,\beta} \mathcal{S}(U_\alpha \cap U_\beta) \quad (3)$$

is exact. Replacing **Ab** with other suitable categories, we have sheaves and presheaves of sets, rings, etc.

Let $f : T \to T'$ be a continuous map. If $\mathcal{S}$ is a presheaf on T, we have the presheaf $f_* \mathcal{S}$ on T' with sections

$$f_* \mathcal{S}(U') := \mathcal{S}(f^{-1}(U')).$$

The restriction maps are given by the obvious formula, and it is easy to see that $f_* \mathcal{S}$ is a sheaf if $\mathcal{S}$ is a sheaf.

Definition 2.1 *A* ringed space *is a pair* $(X, \mathcal{O}_X)$, *where* X *is a topological space, and* $\mathcal{O}_X$ *is a sheaf of rings on* X. *A morphism of ringed spaces* $(X, \mathcal{O}_X) \to (Y, \mathcal{O}_Y)$ *consists of a pair* (f, ϕ), *where* $f : X \to Y$ *is a continuous map, and* $\phi : \mathcal{O}_Y \to f_* \mathcal{O}_X$ *is a homomorphism of sheaves of rings on* Y.

Explicitly, the condition that $\phi : \mathcal{O}_Y \to f_* \mathcal{O}_X$ is a homomorphism of sheaves of rings means that, for each open $V \subset Y$, we have a ring homomorphism

$$\phi(V) : \mathcal{O}_Y(V) \to \mathcal{O}_X(f^{-1}(V)),$$

and for $V' \subset V$, the diagram

$$
\begin{array}{ccc}
\mathcal{O}_Y(V) & \xrightarrow{\phi(V)} & \mathcal{O}_X(f^{-1}(V)) \\
{\scriptstyle \mathrm{res}_{V',V}} \downarrow & & \downarrow {\scriptstyle \mathrm{res}_{f^{-1}(V'), f^{-1}(V)}} \\
\mathcal{O}_Y(V') & \xrightarrow[\phi(V')]{} & \mathcal{O}_X(f^{-1}(V'))
\end{array}
$$

commutes.

Examples 2.2 *(1) Let $\mathcal{T}$ be a topological space, $\mathcal{C}_{\mathcal{T}}$ the presheaf with*

$$\mathcal{C}_{\mathcal{T}}(U) = \{continuous\ functions\ on\ U\}$$

and with the restriction maps the usual restriction of functions. Then $\mathcal{C}_{\mathcal{T}}$ is a sheaf of rings on $\mathcal{T}$. Note that each continuous map $\phi : \mathcal{T} \to \mathcal{S}$ of topological spaces extends to a map of ringed spaces $(\phi, \phi^*) : (\mathcal{T}, \mathcal{C}_{\mathcal{T}}) \to (\mathcal{S}, \mathcal{C}_{\mathcal{S}})$ by

$$\phi^*(g) = g \circ \phi.$$

(2) If $\mathcal{T}$ is a differentiable manifold with underlying topological space denoted $\mathcal{T}_0$, we have as in (1) the sheaf $\mathcal{C}^\infty_{\mathcal{T}_0}$ of C^∞ functions on $\mathcal{T}$, as well as the sheaves $\mathcal{C}^p_{\mathcal{T}_0}$ of C^p functions. These all form sheaves of rings on $\mathcal{T}_0$, giving us the ringed spaces $(\mathcal{T}, \mathcal{C}^p_{\mathcal{T}_0})$, $p = 0, 1, \ldots, \infty$. The differentiable structure on $\mathcal{T}$ is given by the sheaf of rings $\mathcal{C}^\infty_{\mathcal{T}_0}$ on $\mathcal{T}_0$. One can in fact define a differentiable manifold of dimension n as a (paracompact) topological space $\mathcal{T}_0$ with a subsheaf of rings $\mathcal{C}^\infty(\mathcal{T})$ of $\mathcal{C}_{\mathcal{T}}$ such that each point $x \in \mathcal{T}_0$ has a neighborhood U and a homeomorphism ϕ of U with an open subset V of $\mathbb{R}^n$ such that $(\phi, \phi^*) : (U, \mathcal{C}_U) \to (V, \mathcal{C}_V)$ yields an isomorphism of sheaves of rings

$$\phi^* : \mathcal{C}^\infty_V \to \phi_*(\mathcal{C}^\infty(\mathcal{T})_{|U}).$$

Since an open ball in $\mathbb{R}^n$ is diffeomorphic to $\mathbb{R}^n$, we can even assume that $V = \mathbb{R}^n$.

2.2 The Sheaf of Regular Functions on Spec A

In analogy with differentiable manifolds, we want to define a sheaf of rings $\mathcal{O}_X$ on $X = \mathrm{Spec}\,(A)$ so that our functor $A \mapsto \mathrm{Spec}\,(A)$ becomes a faithful functor to the category of ringed spaces. For this, we have seen (lemma 1.15) that the open subsets $X_f := X \setminus V((f))$ form a basis for the topology of X, hence, given $\mathfrak{p} \in \mathrm{Spec}\,(A)$, the open subsets X_f, $f \notin \mathfrak{p}$ form a basis of neighborhoods of $\mathfrak{p}$ in X. Noting that $X_f \cap X_g = X_{fg}$, we see that this basis is closed under finite intersection.

We start our definition of $\mathcal{O}_X$ by setting

$$\mathcal{O}_X(X_f) := A_f;$$

this is justified by proposition 1.16. Suppose that $X_g \subset X_f$. Then $\mathfrak{p} \supset (f) \implies \mathfrak{p} \supset (g)$. Thus, by lemma 1.14, it follows that $\phi_g(f)$ is contained in no proper prime ideal of A_g, hence $\phi_g(f)$ is a unit in A_g. By the universal property of ϕ_f, there is a unique ring homomorphism $\phi_{g,f} : A_f \to A_g$ making the diagram

$$
\begin{array}{ccc}
A & \xrightarrow{\phi_f} & A_f \\
{\scriptstyle \phi_g} \downarrow & \swarrow {\scriptstyle \phi_{g,f}} & \\
A_g & &
\end{array}
$$

By uniqueness, we have

$$\phi_{h,f} = \phi_{h,g} \circ \phi_{g,f} \tag{4}$$

in case $X_h \subset X_g \subset X_f$.

Lemma 2.3 *Let f be in A and let $\{g_\alpha\}$ be a set of elements of A such that $X_f = \cup_\alpha X_{fg_\alpha}$. Then the sequence*

$$0 \to A_f \xrightarrow{\;\Pi\, \phi_{fg_\alpha, f}\;} \prod_\alpha A_{fg_\alpha} \xrightarrow{\;\Pi\, \phi_{fg_\alpha g_\beta, fg_\alpha} - \phi_{fg_\alpha g_\beta, fg_\beta}\;} \prod_{\alpha,\beta} A_{fg_\alpha g_\beta}$$

is exact.

Proof. See [14] or [4]

We thus have the "partially defined" sheaf $\mathcal{O}_X(X_f) = A_f$, satisfying the sheaf axiom for covers consisting of the principal open subsets. Let now $U \subset X$ be an arbitrary open subset. Let $I(U) \subset A$ be the ideal

$$I(U) := \cap_{\mathfrak{p} \notin U}\mathfrak{p},$$

so $U = X \setminus V(I(U))$. Write U as a union of principal open subsets:

$$U = \cup_{f \in I(U)} X_f,$$

and define $\mathcal{O}_X(U)$ as the kernel of the map

$$\prod_{f \in I(U)} \mathcal{O}_X(X_f) \xrightarrow{\;\phi_{fg,f} - \phi_{fg,g}\;} \prod_{f,g \in I(U)} \mathcal{O}_X(X_{fg}).$$

If we have $V \subset U$, then $I(V) \subset I(U)$, and so we have the projections

$$\pi_{V,U} : \prod_{f \in I(U)} \mathcal{O}_X(X_f) \to \prod_{f \in I(V)} \mathcal{O}_X(X_f)$$

$$\pi'_{V,U} : \prod_{f,g \in I(U)} \mathcal{O}_X(X_{fg}) \to \prod_{f,g \in I(V)} \mathcal{O}_X(X_{fg})$$

These in turn induce the map

$$\mathrm{res}_{V,U} : \mathcal{O}_X(U) \to \mathcal{O}_X(V)$$

satisfying $\mathrm{res}_{W,V} \circ \mathrm{res}_{V,U} = \mathrm{res}_{W,U}$ for $W \subset V \subset U$.

We now have two definitions of $\mathcal{O}_X(U)$ in case $U = X_f$, but by lemma 2.3, these two agree. It remains to check the sheaf axiom for an arbitrary cover of an arbitrary open $U \subset X$, but this follows formally from lemma 2.3. Thus, we have the sheaf of rings $\mathcal{O}_X$ on X.

Let $\psi : A \to B$ be a homomorphism of commutative rings, giving us the continuous map $\hat{\psi} : Y := \mathrm{Spec}\,(B) \to X := \mathrm{Spec}\,(A)$. Take $f \in B$, $g \in A$, and suppose that $\hat{\psi}(Y_f) \subset X_g$. This says that, if $\mathfrak{p}$ is a prime idea in B with $f \notin \mathfrak{p}$, then $\psi(g) \notin \mathfrak{p}$. This implies that $\phi_f(\psi(g))$ is in no prime ideal of B_f, hence $\phi_f(\psi(g))$ is a unit in B_f. By the universal property of $\phi_g : A \to A_g$, there is a unique ring homomorphism $\psi_{f,g} : A_g \to B_f$ making the diagram

$$
\begin{array}{ccc}
A & \xrightarrow{\psi} & B \\
\phi_g \downarrow & & \downarrow \phi_f \\
A_g & \xrightarrow{\psi_{f,g}} & B_f
\end{array}
$$

commute. One easily checks that the $\psi_{f,g}$ fit together to define the map of sheaves

$$\tilde{\psi} : \mathcal{O}_X \to \hat{\psi}_* \mathcal{O}_Y,$$

giving the map of ringed spaces

$$(\hat{\psi}, \tilde{\psi}) : (Y, \mathcal{O}_Y) \to (X, \mathcal{O}_X).$$

The functoriality

$$(\widehat{\psi_1 \circ \psi_2}, \widetilde{\psi_1 \circ \psi_2}) = (\hat{\psi}_2, \tilde{\psi}_2) \circ (\hat{\psi}_1, \tilde{\psi}_1)$$

is also easy to check.

Thus, we have the contravariant functor Spec from commutative rings to ringed spaces. Since $\mathcal{O}_X(X) = A$ if $X = \operatorname{Spec} A$, Spec is clearly a faithful functor.

2.3 Local Rings and Stalks

Recall that a commutative ring $\mathcal{O}$ is called a *local* ring if $\mathcal{O}$ has a unique maximal ideal $\mathfrak{m}$. The field $\mathfrak{k} := \mathcal{O}/\mathfrak{m}$ is the *residue field* of $\mathcal{O}$. A homomorphism $f : \mathcal{O} \to \mathcal{O}'$ of local rings is called a local homomorphism if $f(\mathfrak{m}) \subset \mathfrak{m}'$.

Example 2.4 *Let A be a commutative ring, $\mathfrak{p} \subset A$ a proper prime ideal. Let $S = A \backslash \mathfrak{p}$; S is then a multiplicatively closed subset of A containing 1. We set $A_{\mathfrak{p}} := S^{-1}A$, and write $\mathfrak{p}A_{\mathfrak{p}}$ for the ideal generated by $\phi_S(\mathfrak{p})$.*

We claim that $A_{\mathfrak{p}}$ is a local ring with maximal ideal $\mathfrak{p}A_{\mathfrak{p}}$. Indeed, each proper prime ideal of $A_{\mathfrak{p}}$ is the ideal generated by $\phi_S(\mathfrak{q})$, for $\mathfrak{q}$ some prime ideal of A with $\mathfrak{q} \cap S = \emptyset$. As this is equivalent to $\mathfrak{q} \subset \mathfrak{p}$, we find that $\mathfrak{p}A_{\mathfrak{p}}$ is indeed the unique maximal ideal of $A_{\mathfrak{p}}$.

Definition 2.5 *Let $\mathcal{F}$ be a sheaf (of sets, abelian, rings, etc.) on a topological space X. The stalk of $\mathcal{F}$ at x, written $\mathcal{F}_x$, is the direct limit*

$$\mathcal{F}_x := \varinjlim_{x \in U} \mathcal{F}(U).$$

Note that, if $(f, \phi) : (Y, \mathcal{O}_Y) \to (X, \mathcal{O}_X)$ is a morphism of ringed spaces, and we take $y \in Y$, then $\phi : \mathcal{O}_X \to f_* \mathcal{O}_Y$ induces the homomorphism of stalks

$$\phi_y : \mathcal{O}_{X, f(y)} \to \mathcal{O}_{Y, y}.$$

Lemma 2.6 *Let A be a commutative ring, $(X, \mathcal{O}_X) = \operatorname{Spec}(A)$. Then for $\mathfrak{p} \in \operatorname{Spec}(A)$, we have*

$$\mathcal{O}_{X, \mathfrak{p}} = A_{\mathfrak{p}}.$$

Proof. We may use the principal open subsets X_f, $f \notin \mathfrak{p}$, to define the stalk $\mathcal{O}_{X, \mathfrak{p}}$:

$$
\begin{aligned}
\mathcal{O}_{X, \mathfrak{p}} &= \varinjlim_{f \notin \mathfrak{p}} \mathcal{O}_X(X_f) \\
&= \varinjlim_{f \notin \mathfrak{p}} A_f \\
&= (A \setminus \mathfrak{p})^{-1} A \\
&= A_{\mathfrak{p}}.
\end{aligned}
$$

Thus, the ringed spaces of the form $(X, \mathcal{O}_X) = \operatorname{Spec}(A)$ are special, in that the stalks of the sheaf $\mathcal{O}_X$ are all *local* rings. The morphisms $(\hat{\psi}, \tilde{\psi})$ coming from ring homomorphisms $\psi : A \to B$ are also special:

Lemma 2.7 *Let $(X, \mathcal{O}_X) = \operatorname{Spec} A$, $(Y, \mathcal{O}_Y) = \operatorname{Spec} B$, and let $\psi : A \to B$ be a ring homomorphism. Take $y \in Y$. Then*

$$\tilde{\psi}_y : \mathcal{O}_{X, \hat{\psi}(y)} \to \mathcal{O}_{Y, y}$$

is a local homomorphism.

Proof. In fact, if y is the prime ideal $\mathfrak{p} \subset B$, then $\hat{\psi}(y)$ is the prime ideal $\mathfrak{q} := \psi^{-1}(\mathfrak{p})$, and $\tilde{\psi}_y$ is just the ring homomorphism

$$\psi_{\mathfrak{p}} : A_{\mathfrak{q}} \to B_{\mathfrak{p}}$$

induced by ψ, using the universal property of localization. Since

$$\psi(\psi^{-1}(\mathfrak{p})) \subset \mathfrak{p},$$

$\psi_{\mathfrak{p}}$ is a local homomorphism.

3 The Category of Schemes

As we have seen, one can view a differentiable manifold of dimension n as a ringed space that is locally the same as $(\mathbb{R}^n, C_{\mathbb{R}^n}^\infty)$. Grothendieck defined a *scheme* in roughly the same way, with the important difference that, rather than one local model $\mathbb{R}^n$ in each dimension, one needs to use *all* the ringed spaces $\operatorname{Spec} A$ for the local models.

3.1 Objects and Morphisms

Definition 3.1 *A* scheme *is a ringed space* $(X, \mathcal{O}_X)$ *which is locally* Spec *of a ring, i.e., for each point* $x \in X$, *there is an open neighborhood* U *of* x *and a commutative ring* A *such that* $(U, \mathcal{O}_U)$ *is isomorphic to* Spec A *as a ringed space, where* $\mathcal{O}_U$ *denotes the restriction of* $\mathcal{O}_X$ *to a sheaf of rings on* U.

A morphism of schemes *$f : (X, \mathcal{O}_X) \to (Y, \mathcal{O}_Y)$ is a morphism of ringed spaces which is locally of the form $(\hat{\psi}, \tilde{\psi})$ for some homomorphism of commutative rings $\psi : A \to B$, that is, for each $x \in X$, there are neighborhoods U of x and V of $f(x)$ such that f restricts to a map $f_U : (U, \mathcal{O}_U) \to (V, \mathcal{O}_V)$, a homomorphism of commutative rings $\psi : A \to B$ and isomorphisms of ringed spaces*

$$(U, \mathcal{O}_U) \xrightarrow{g} \operatorname{Spec} B; \quad (V, \mathcal{O}_V) \xrightarrow{h} \operatorname{Spec} A$$

making the diagram

$$
\begin{array}{ccc}
(U, \mathcal{O}_U) & \xrightarrow{\ f_U\ } & (V, \mathcal{O}_V) \\
{\scriptstyle g}\downarrow & & \downarrow{\scriptstyle h} \\
\operatorname{Spec} B & \xrightarrow[(\hat{\psi}, \tilde{\psi})]{} & \operatorname{Spec} A
\end{array}
$$

commute.

Note that, as the functor Spec is faithful, the homomorphism ψ is uniquely determined by f once we choose isomorphisms g and h. We will see below that the functor Spec is *fully faithful*, which implies that the isomorphism $(U, \mathcal{O}_U) \to \operatorname{Spec} A$ required in the first part of Definition 3.1 is unique up to unique isomorphism of rings $A \to A'$. Thus, the data in the second part of the definition are uniquely determined (up to unique isomorphism) once one fixes the open neighborhoods U and V.

Definition 3.2 *A scheme $(X, \mathcal{O}_X)$ isomorphic to* Spec (A) *for some commutative ring A is called an* affine *scheme. An open subset $U \subset X$ such that $(U, \mathcal{O}_U)$ is an affine scheme is called an* affine open subset *of X.*

If $U \subset X$ is an affine open subset, then $(U, \mathcal{O}_U) = \operatorname{Spec} A$, where $A = \mathcal{O}_X(U)$.

Let **Sch** denote the category of schemes, and **Aff** $\subset$ **Sch** the full subcategory of affine schemes. Sending A to Spec A thus defines the functor

$$\operatorname{Spec} : \mathbf{Rings}^{\mathrm{op}} \to \mathbf{Aff}$$

Lemma 3.3 Spec *is an equivalence of categories; in particular*

$$\operatorname{Hom}_{\mathbf{Rings}}(A, B) \cong \operatorname{Hom}_{\mathbf{Sch}}(\operatorname{Spec} B, \operatorname{Spec} A).$$

Proof. We have already seen that Spec is a faithful functor, so it remains to see that Spec is full, i.e., a morphism of affine schemes $f : \operatorname{Spec} B \to \operatorname{Spec} A$ arises from a homomorphism of rings.

For this, write $\operatorname{Spec} A = (X, \mathcal{O}_X)$, $\operatorname{Spec} B = (Y, \mathcal{O}_Y)$. f gives us the ring homomorphism

$$f^*(Y, X) : \mathcal{O}_X(X) \to \mathcal{O}_Y(Y),$$

i.e., a ring homomorphism $\psi : A \to B$, so we need to see that $f = (\hat{\psi}, \tilde{\psi})$.

Take $y \in Y$ and let $x = f(y)$. Then as f locally of the form $(\hat{\phi}, \tilde{\phi})$, the homomorphism $f_y^* : \mathcal{O}_{Y,y} \to \mathcal{O}_{X,x}$ is local and $(f_y^*)^{-1}(\mathfrak{m}_x) = \mathfrak{m}_y$, where $\mathfrak{m}_x$ and $\mathfrak{m}_y$ are the respective maximal ideals.

Suppose x is the prime ideal $\mathfrak{p} \subset A$ and y is the prime ideal $\mathfrak{q} \subset B$. We thus have $\mathcal{O}_{Y,y} = B_{\mathfrak{q}}$, $\mathcal{O}_{X,x} = A_{\mathfrak{p}}$, and the diagram

$$
\begin{array}{ccc}
A & \xrightarrow{\psi} & B \\
\phi_{\mathfrak{p}} \downarrow & & \downarrow \phi_{\mathfrak{q}} \\
A_{\mathfrak{p}} & \xrightarrow{f_y^*} & B_{\mathfrak{q}}
\end{array}
$$

commutes. Since f_y^* is local, it follows that $\psi^{-1}(\mathfrak{q}) \subset \mathfrak{p}$; as $(f_y^*)^{-1}(\mathfrak{m}_x) = \mathfrak{m}_y$ and $\phi_{\mathfrak{p}}$ induces a bijection between the prime ideals of A contained in $\mathfrak{p}$ and the prime ideals of $A_{\mathfrak{p}}$, it follows that $\mathfrak{p} = \psi^{-1}(\mathfrak{q})$. Thus, as maps of topological spaces, f and $\hat{\psi}$ agree.

We note that the map $\psi_{\mathfrak{p},\mathfrak{q}} : A_{\mathfrak{p}} \to A_{\mathfrak{q}}$ induced by localizing ψ is the unique local homomorphism ρ making the diagram

$$
\begin{array}{ccc}
A & \xrightarrow{\psi} & B \\
\phi_{\mathfrak{p}} \downarrow & & \downarrow \phi_{\mathfrak{q}} \\
A_{\mathfrak{p}} & \xrightarrow{\rho} & B_{\mathfrak{q}}
\end{array}
$$

commute. Thus $f_y^* = \psi_{\mathfrak{p},\mathfrak{q}}$. Now, for $U \subset X$ open, the map

$$\mathcal{O}_X(U) \to \prod_{x \in U} \mathcal{O}_{X,x}$$

is injective, and similarly for open subsets V of Y. Thus $f^* = \tilde{\psi}$, completing the proof.

Remark 3.4 *It follows from the proof of lemma 3.3 that one can replace the condition in Definition 3.1 defining a morphism of schemes with the following:*

If $(X, \mathcal{O}_X)$ and $(Y, \mathcal{O}_Y)$ are schemes, a morphism of ringed spaces $(f, \phi) : (X, \mathcal{O}_X) \to (Y, \mathcal{O}_Y)$ is a morphism of schemes if for each $x \in X$, the map $\phi_x : \mathcal{O}_{Y,f(x)} \to \mathcal{O}_{X,x}$ is a local homomorphism.

Remark 3.5 *Let $(X, \mathcal{O}_X)$, and $(Y, \mathcal{O}_Y)$ be arbitrary schemes. We have the rings $A := \mathcal{O}_X(X)$ and $B := \mathcal{O}_Y(Y)$, and each map of schemes $f : (Y, \mathcal{O}_Y) \to (X, \mathcal{O}_X)$ induces the ring homomorphism $f^*(X) : A \to B$, giving us the functor*

$$\Gamma : \mathbf{Sch} \to \mathbf{Rings}^{\mathrm{op}}.$$

We have seen above that $\Gamma \circ \mathrm{Spec} = \mathrm{id}_{\mathbf{Rings}}$, and that the restriction of Γ to $\mathbf{Aff}$ is the inverse to Spec. More generally, suppose that X is affine, $(X, \mathcal{O}_X) = \mathrm{Spec}\, A$. Then

$$\Gamma : \mathrm{Hom}_{\mathbf{Sch}}(Y, X) \to \mathrm{Hom}_{\mathbf{Rings}}(A, B)$$

is an isomorphism. This is not *the case in general for non-affine X.*

It is definitely *not* the case that sending a ring A to the topological space $|\mathrm{Spec}\, A|$ underlying the ringed space $\mathrm{Spec}\, A$ distinguishes non-isomorphic rings from each other; this functor is not even faithful.

Examples 3.6 *(1) For a field F, $\mathrm{Spec}\, F$ is a single point with (constant) sheaf of functions F. Thus, a field extension $\phi : F \to L$ induces the identity map $|\mathrm{Spec}\, L| \to |\mathrm{Spec}\, F|$ and the map ϕ on the sheaves of rings.*

(2) Let $A = k[\epsilon]/(\epsilon^2)$. $(\epsilon) \subset A$ is the unique prime ideal, so $|\mathrm{Spec}\, A|$ is a single point and the reduction map $A \to k$ sending ϵ to zero induces the identity map $|\mathrm{Spec}\, A| \to |\mathrm{Spec}\, k|$. More generally, for a commutative ring A, let $\mathcal{N} \subset A$ be the ideal of nilpotent elements. Then $A \to A/\mathcal{N}$ induces a homeomorphism $|\mathrm{Spec}\, A| \to |\mathrm{Spec}\, A/\mathcal{N}|$. We will say more about this phenomenon in section 4.2.

3.2 Gluing Constructions

We have seen the analogy between differentiable manifolds and schemes, in that a scheme is a locally affine ringed space, while a differentiable manifold is a ringed space, locally isomorphic to $(\mathbb{R}^n, \mathcal{C}_{\mathbb{R}^n}^\infty)$. As for manifolds, one can construct a scheme by *gluing*: Let U_α be a collection of schemes, together with open subschemes $U_{\alpha,\beta} \subset U_\alpha$, and isomorphisms

$$g_{\beta,\alpha} : U_{\alpha,\beta} \to U_{\beta,\alpha}$$

satisfying the cocycle condition

$$g_{\beta,\alpha}^{-1}(U_{\beta,\alpha} \cap U_{\beta,\gamma}) = U_{\alpha,\beta} \cap U_{\alpha,\gamma}$$

$$g_{\gamma,\alpha} = g_{\gamma,\beta} \circ g_{\beta,\alpha} \text{ on } U_{\alpha,\beta} \cap U_{\alpha,\gamma}.$$

This allows one to define the underlying topological space of the glued scheme X by gluing the underlying spaces of the schemes U_α; the structure sheaf $\mathcal{O}_X$ is constructed by gluing the structure sheaves $\mathcal{O}_{U_\alpha}$. If the U_α and the open subschemes $U_{\alpha,\beta}$ are all affine, the entire structure is defined via commutative rings and ring homomorphisms.

Remark 3.7 *Throughout the text, we will define various properties along these lines by requiring certain conditions hold on some affine open cover of X. It is usually the case that these defining conditions then hold on every affine open subscheme (see below) of X, and we will use this fact without further explicit mention.*

3.3 Open and Closed Subschemes

Let $(X, \mathcal{O}_X)$ be a scheme. An *open subscheme* of X is a scheme of the form $(U, \mathcal{O}_U)$, where $U \subset X$ is an open subspace. A morphism of schemes $j : (V, \mathcal{O}_V) \to (X, \mathcal{O}_X)$ which gives rise to an isomorphism $(V, \mathcal{O}_V) \cong (U, \mathcal{O}_U)$ for some open subscheme $(U, \mathcal{O}_U)$ of $(X, \mathcal{O}_X)$ is called an *open immersion*.

Closed subschemes are a little less straightforward. We first define the notion of a *sheaf of ideals*. Let $(X, \mathcal{O}_X)$ be a ringed space. We have the category of $\mathcal{O}_X$-modules: an $\mathcal{O}_X$-module is a sheaf of abelian groups $\mathcal{M}$ on X, together with a map of sheaves

$$\mathcal{O}_X \times \mathcal{M} \to \mathcal{M}$$

which is associative and unital (in the obvious sense). Morphisms are maps of sheaves respecting the multiplication. Now suppose that $(X, \mathcal{O}_X) = \operatorname{Spec} A$ is an affine scheme, and $I \subset A$ is an ideal. For each $f \in A$, we have the ideal $I_f \subset A_f$, being the ideal generated by $\phi_f(I)$. These patch together to form an $\mathcal{O}_X$-submodule of $\mathcal{O}_X$, called the *ideal sheaf generated by I*, and denoted $\tilde{I}$. In general, if X is a scheme, we call an $\mathcal{O}_X$-submodule $\mathcal{I}$ of $\mathcal{O}_X$ an *ideal sheaf* if $\mathcal{I}$ is locally of the form $\tilde{I}$ for some ideal $I \subset \mathcal{O}_X(U)$, $U \subset X$ affine.

Let $\mathcal{I}$ be an ideal sheaf. We may form the sheaf of rings $\mathcal{O}_X / \mathcal{I}$ on X. The *support* of $\mathcal{O}_X / \mathcal{I}$ is the (closed) subset of X consisting of those x with $(\mathcal{O}_X / \mathcal{I})_x \neq 0$. Letting $i : Z \to X$ be the inclusion of the support of $\mathcal{O}_X / \mathcal{I}$, we have the ringed space $(Z, \mathcal{O}_X / \mathcal{I})$ and the morphism of ringed spaces $(i, \pi) : (Z, \mathcal{O}_X / \mathcal{I}) \to (X, \mathcal{O}_X)$, where π is given by the surjection $\mathcal{O}_X \to i_* \mathcal{O}_X / \mathcal{I}$. If we take an affine open subscheme $U = \operatorname{Spec} A$ of X and an ideal $I \subset A$ for which $\mathcal{I}_{|U} = \tilde{I}$, then one has

$$Z \cap U = V(I).$$

We call $(Z, \mathcal{O}_X / \mathcal{I})$ a *closed subscheme* of $(X, \mathcal{O}_X)$. More generally, let $i : (W, \mathcal{O}_W) \to (X, \mathcal{O}_X)$ a morphism of schemes such that

1. $i : W \to X$ gives a homeomorphism of W with a closed subset Z of X.
2. $i^* : \mathcal{O}_X \to i_* \mathcal{O}_W$ is surjective, with kernel an ideal sheaf.

Then we call i a *closed immersion*.

If $\mathcal{I} \subset \mathcal{O}_X$ is the ideal sheaf associated to a closed subscheme $(Z, \mathcal{O}_Z)$, we call $\mathcal{I}$ the *ideal sheaf* of $(Z, \mathcal{O}_Z)$, and write $\mathcal{I} = \mathcal{I}_Z$.

Examples 3.8 *(1) Let $(X, \mathcal{O}_X) = \operatorname{Spec} A$, and $I \subset A$ an ideal, giving us the ideal sheaf $\tilde{I} \subset \mathcal{O}_X$, and the closed subscheme $i : (Z, \mathcal{O}_Z) \to (X, \mathcal{O}_X)$. Then*

$(Z, \mathcal{O}_Z)$ is affine, $(Z, \mathcal{O}_Z) = \operatorname{Spec} A/I$, and $i = (\hat{\pi}, \tilde{\pi})$, where $\pi : A \to A/I$ is the canonical surjection. Conversely, if $(Z, \mathcal{O}_Z)$ is a closed subscheme of $\operatorname{Spec} A$ with ideal sheaf $\mathcal{I}_Z$, then $(Z, \mathcal{O}_Z)$ is affine, $(Z, \mathcal{O}_Z) = \operatorname{Spec} A/I$ (as closed subscheme), where $I = \mathcal{I}_Z(X)$.

(2) The closed subschemes Z of $\mathbb{A}^1_F$, F a field, with $|Z|$ a single point, are of the form $\operatorname{Spec} F[x]/(f^n)$, where $f \in F[x]$ is an irreducible polynomial, and $n \geq 1$ is an integer. Two such are equal (as subschemes)

$$\operatorname{Spec} F[x]/(f^n) = \operatorname{Spec} F[x]/(g^m)$$

if and only if $n = m$ and $f = c \cdot g$ for some non-zero $c \in F$. However, these two are isomorphic as schemes exactly when the fields $F[x]/(f)$ and $F[x]/(g)$ are isomorphic and $n = m$.

(3) The closed subschemes Z of $\operatorname{Spec} \mathbb{Z}$ with $|Z|$ a single point are those of the form $\operatorname{Spec} \mathbb{Z}/(p^n)$, $n \geq 1$ an integer, $p > 0$ a prime number. These are all distinct, both as closed subschemes of $\operatorname{Spec} \mathbb{Z}$ and as schemes.

3.4 Fiber Products

An important property of the category **Sch** is the existence of a (categorical) fiber product $Y \times_X Z$ for each pair of morphisms $f : Y \to X$, $g : Z \to X$. We sketch the construction.

First consider the case of affine $X = \operatorname{Spec} A$, $Y = \operatorname{Spec} B$ and $Z = \operatorname{Spec} C$. We thus have ring homomorphisms $f^* : A \to B$, $g^* : A \to C$ with $f = (\hat{f}^*, \tilde{f}^*)$ and similarly for g. The maps $B \to B \otimes_A C$, $C \to B \otimes_A C$ defined by $b \mapsto b \otimes 1$, $c \mapsto 1 \otimes c$ give us the commutative diagram of rings

$$
\begin{array}{ccc}
B \otimes_A C & \longleftarrow & B \\
\uparrow & & \uparrow{\scriptstyle f^*} \\
C & \xleftarrow{\;g^*\;} & A.
\end{array}
$$

It is not hard to see that this exhibits $B \otimes_A C$ as the categorical coproduct of B and C over A (in **Rings**), and thus applying Spec yields the fiber product diagram

$$
\begin{array}{ccc}
\operatorname{Spec}(B \otimes_A C) & \longrightarrow & Y \\
\downarrow & & \downarrow{\scriptstyle f} \\
Z & \xrightarrow{\;g\;} & X
\end{array}
$$

in **Aff** $=$ **Rings**^{op}. By remark 3.5, this diagram is a fiber product diagram in **Sch** as well.

Now suppose that X, Y and Z are arbitrary schemes. We can cover X, Y and Z by affine open subschemes X_α, Y_α, Z_α such that the maps f and g restrict to $f_\alpha : Y_\alpha \to X_\alpha$, $g_\alpha : Z_\alpha \to X_\alpha$ (we may have $X_\alpha = X_\beta$ for $\alpha \neq \beta$). We thus have the fiber products $Y_\alpha \times_{X_\alpha} Z_\alpha$ for each α. The universal property of fiber products together with the gluing data for the individual covers $\{X_\alpha\}$, $\{Y_\alpha\}$, $\{Z_\alpha\}$ defines gluing data for the pieces $Y_\alpha \times_{X_\alpha} Z_\alpha$, which forms the desired categorical fiber product $Y \times_X Z$.

Examples 3.9 *(1)* $\mathbb{A}_F^n \times_{\operatorname{Spec} F} \mathbb{A}_F^m \cong \mathbb{A}_F^{n+m}$ *since*

$$F[x_1, \ldots, x_n] \otimes_F F[y_1, \ldots, y_m] \cong F[z_1, \ldots, z_{n+m}].$$

(2) Let $\phi : F[x] \to F[s,t]$ be the homomorphism over F sending x to $s^2 - x$, giving dually the map of schemes $f : \mathbb{A}_F^2 \to \mathbb{A}_F^1$. The fiber product $Y \times_X Z$, with $Y = Z = \mathbb{A}_F^2$, $X = \mathbb{A}_F^1$ and $Y \to X$, $Z \to X$ both the map f, is just $\operatorname{Spec} F[s,t] \otimes_{F[x]} F[s,t]$. The ring $F[s,t] \otimes_{F[x]} F[s,t]$ is the quotient of $F[s,t] \otimes_F F[s,t] \cong F[x_1, x_2, x_3, x_4]$ by the relation $x_1^2 - x_2 = x_3^2 - x_4$, which makes the fiber product $Y \times_X Z$ isomorphic to the closed subscheme of $\mathbb{A}_F^4$ defined by the ideal $(x_1^2 - x_2 - x_3^2 + x_4)$.

Example 3.10 (The fibers of a morphism) *Let X be a scheme, $x \in |X|$ a point, $k(x)$ the residue field of the local ring $\mathcal{O}_{X,x}$. The inclusion $x \to |X|$ together with the residue map $\mathcal{O}_{X,x} \to k(x)$ defines the morphism of schemes $i_x : \operatorname{Spec} k(x) \to X$. If $f : Y \to X$ is a morphism, we set*

$$f^{-1}(x) := \operatorname{Spec} k(x) \times_X Y.$$

Via the first projection, $f^{-1}(x)$ is a scheme over $k(x)$, called the fiber of f over x.

4 Schemes and Morphisms

In practice, one restricts attention to various special types of schemes and morphisms. In this section, we describe the most important of these.

For a scheme X, we write $|X|$ for the underlying topological space of X, and $\mathcal{O}_X$ for the structure sheaf of rings on $|X|$. For a morphism $f : Y \to X$ of schemes, we usually write $f : |Y| \to |X|$ for the map of underlying spaces (sometimes $|f|$ if necessary) and $f^* : \mathcal{O}_X \to f_*\mathcal{O}_Y$ for the map of sheaves of rings.

4.1 Noetherian Schemes

We amplify our discussion of noetherian rings and noetherian spaces begun in section 1.2, where we defined a noetherian space and a noetherian ring (definition 1.3).

Recall that a commutative ring A is noetherian if the following equivalent conditions are satisfied:

1. Every increasing sequence of ideals in A

$$I_0 \subset I_1 \subset \ldots \subset I_n \subset \ldots$$

   is eventually constant.
2. Let M be a finitely generated A-module (i.e., there exist elements $m_1, \ldots, m_n \in M$ such that each element of M is of the form $\sum_{i=1}^{n} a_i m_i$ with the $a_i \in A$). Then every increasing sequence of submodules of M

$$N_0 \subset N_1 \subset \ldots \subset N_n \subset \ldots$$

   is eventually constant.
3. Let M be a finitely generated A-module, $N \subset M$ a submodule. Then N is finitely generated as an A-module.
4. Let I be an ideal in A. Then I is a finitely generated ideal.

If A is a noetherian ring, then the topological space $|\operatorname{Spec} A|$ is a noetherian topological space (but not conversely).

We call a scheme X noetherian if $|X|$ is noetherian, and X admits an affine cover, $X = \cup_\alpha \operatorname{Spec} A_\alpha$ with each A_α a noetherian ring. We can take the cover to have finitely many elements. If X is a noetherian scheme, so is each open or closed subscheme of X. In particular, $\operatorname{Spec} A$ is a noetherian scheme for each noetherian ring A.

Examples 4.1 *(1) A field k is clearly a noetherian ring. Recall that a principal ideal domain (PID) is a ring A such that all ideals are generated by a single element; clearly a PID is noetherian, in particular, $\mathbb{Z}$ is noetherian.*

(2) The Hilbert basis theorem *states that, if A is noetherian, so is the polynomial ring $A[x]$. Thus, the polynomial rings $k[x_1, \ldots, x_n]$ (k a field) and $\mathbb{Z}[x_1, \ldots, x_n]$ are noetherian. If A is noetherian, so is A/I for each ideal I, thus, every quotient of $A[x_1, \ldots, x_n]$ (that is, every A-algebra that is finitely generated as an A-algebra) is noetherian. This applies to, e.g., a ring of integers in a number field, or rings of the form $k[x_1, \ldots, x_n]/I$.*

(3) Let A be a ring. Affine n-space over A is the affine scheme

$$\mathbb{A}_A^n := \operatorname{Spec} A[x_1, \ldots, x_n].$$

If A is a noetherian ring, then $\mathbb{A}_A^n$ is a noetherian scheme.

4.2 Irreducible Schemes, Reduced Schemes and Generic Points

Let X be a topological space. X is called *irreducible* if X is not the union of two proper closed subsets; equivalently, each non-empty open subspace of X is dense. We have already seen that a noetherian topological space X is uniquely a finite union of irreducible closed subspaces

$$X = X_0 \cup \ldots \cup X_N$$

where no X_i contains X_j for $i \neq j$. We call a scheme X *irreducible* if $|X|$ is an irreducible topological space.

Example 4.2 *Let A be an integral domain. Then $\operatorname{Spec} A$ is irreducible. Indeed, since A is a domain, (0) is a prime ideal, and as every prime ideal contains (0), $\operatorname{Spec} A$ is the closure of the singleton set $\{(0)\}$, hence irreducible. The point $x_{\text{gen}} \in \operatorname{Spec} A$ corresponding to the prime ideal (0) is the* generic point *of $\operatorname{Spec} A$.*

An element x in a ring A is called *nilpotent* if $x^n = 0$ for some n; the set of nilpotent elements in a commutative ring A form an ideal, $rad(A)$, called the *radical* of A. Note that $rad(A)$ is just the radial $\sqrt{(0)}$ of the zero-ideal. A ring A is *reduced* if $rad(A) = \{0\}$; clearly $A_{\text{red}} := A/rad(A)$ is the maximal reduced quotient of A. As one clearly has $rad(A) \subset \mathfrak{p}$ for every prime ideal $\mathfrak{p}$ of A, the quotient map $A \to A_{\text{red}}$ induces a homeomorphism $\operatorname{Spec} A_{\text{red}} \to \operatorname{Spec} A$.

A scheme X is called *reduced* if for each open subset $U \subset |X|$, the ring $\mathcal{O}_X(U)$ contains no non-zero nilpotent elements. A reduced, irreducible scheme is called *integral*. If $X = \operatorname{Spec} A$ is affine, then X is an integral scheme if and only if A is a domain.

For a scheme X, we have the sheaf $rad_X \subset \mathcal{O}_X$ with $rad_X(U)$ the set of nilpotent elements of $\mathcal{O}_X(U)$. rad_X is a sheaf of ideals; we let X_{red} be the closed subscheme of X with structure sheaf $\mathcal{O}_X/rad_X$. The closed immersion $X_{\text{red}} \to X$ is a homeomorphism since this is the case for affine X. Thus each scheme X has a canonical reduced closed subscheme X_{red} homeomorphic to X. More generally, if $Z \subset |X|$ is a closed subset, there is a unique sheaf of ideals $\mathcal{I}_Z$ such that the closed subscheme W of X defined by $\mathcal{I}_Z$ has underlying topological space Z, and W is reduced. We usually write Z for this closed subscheme, and say that we give Z the reduced subscheme structure.

Lemma 4.3 *Let X be a non-empty irreducible scheme. Then $|X|$ has a unique point x_{gen} with $|X|$ the closure of x_{gen}. x_{gen} is called the* generic point *of X*

Proof. Replacing X with X_{red}, we may assume that X is integral. If $U \subset X$ is an affine open subscheme, then U is dense in X and U is integral, so $U = \operatorname{Spec} A$ with A a domain. We have already seen that $|U|$ satisfies the lemma. If $u \in |U|$ is the generic point, then the closure of u in $|X|$ is $|X|$; uniqueness of x_{gen} follows from the uniqueness of u_{gen} and denseness of U.

If X is noetherian, then $|X|$ has finitely many irreducible components: $|X| = \cup_{i=1}^N X_i$, without containment relations among the X_i. The X_i (with the reduced subscheme structure) are called the *irreducible components* of X, and the generic points $x_1, \ldots, x_N$ of the components X_i are called the *generic points* of X.

Definition 4.4 *Let X be a noetherian scheme with generic points $x_1, \ldots, x_N$. The* ring of rational functions *on X is the ring*

$$k(X) := \prod_{i=1}^{N} \mathcal{O}_{X,x_i}.$$

If X is integral, then $k(X)$ is a field, called the field of rational functions on X.

4.3 Separated Schemes and Morphisms

We have already seen that, except for trivial cases, the topological space $|X|$ underlying a scheme X is not Hausdorff. So, to replace the usual separation axioms, we have the following condition: a morphism of schemes $f : X \to Y$ is called *separated* if the diagonal inclusion $X \to X \times_Y X$ has image $\delta(|X|)$ a closed subset of $|X \times_Y X|$. Noting that every scheme has a unique morphism to $\operatorname{Spec} \mathbb{Z}$, we call X a separated scheme if X is separated over $\operatorname{Spec} \mathbb{Z}$, i.e., the diagonal X in $X \times_{\operatorname{Spec} \mathbb{Z}} X$ is closed.

Separation has the following basic properties:

Proposition 4.5 *Let $X \xrightarrow{f} Y \xrightarrow{g} Z$ be morphisms of schemes.*

1. *If gf is separated, then f is separated.*
2. *If f is separated and g is separated, the gf is separated.*
3. *if f is separated and $h : W \to Y$ is an arbitrary morphism, then the projection $X \times_Y W \to W$ is separated.*

In addition:

Proposition 4.6 *Every affine scheme is separated.*

Proof. Let $X = \operatorname{Spec} A$. The diagonal inclusion $\delta : X \to X \times_{\operatorname{Spec} \mathbb{Z}} X$ arises from the dual diagram of rings

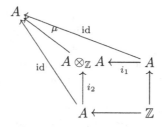

where $i_1(a) = a \otimes 1$, $i_2(a) = 1 \otimes a$, and the maps $\mathbb{Z} \to A$ are the canonical ones. Thus μ is the multiplication map, hence surjective. Letting $I \subset A \otimes_{\mathbb{Z}} A$ be the kernel of μ (in fact I is the ideal generated by elements $a \otimes 1 - 1 \otimes a$), we see that δ is a closed immersion with image $\operatorname{Spec} A \otimes_{\mathbb{Z}} A/I$.

The point of using separated schemes is that this forces the condition that two morphisms $f, g : X \to Y$ be the same to be a *closed* subscheme of X. Indeed, the equalizer of f and g is the pull-back of the diagonal $\delta_Y \subset Y \times Y$ via the map $(f, g) : X \to Y \times Y$, so this equalizer is closed if δ_Y is closed in $Y \times Y$. Another nice consequence is

Proposition 4.7 *Let X be a separated scheme, U and V affine open subschemes. Then $U \cap V$ is also affine.*

Proof. We have the fiber product diagram

$$
\begin{array}{ccc}
U \cap V & \xrightarrow{\ i_{U \cap V}\ } & X \\
\left\downarrow{\scriptstyle \delta'}\right. & & \left\downarrow{\scriptstyle \delta_X}\right. \\
U \times V & \xrightarrow[\ i_U \times i_V\ ]{} & X \times X
\end{array}
$$

identifying $U \cap V$ with $(U \times V) \times_{X \times X} X$. If $U = \operatorname{Spec} A$ and $V = \operatorname{Spec} B$, then $U \times V = \operatorname{Spec} A \otimes B$ is affine; since δ_X is a closed immersion, so is δ'. Thus $U \cap V$ is isomorphic to a closed subscheme of the affine scheme $\operatorname{Spec} A \otimes B$, hence

$$
U \cap V = \operatorname{Spec} A \otimes B / I
$$

for some ideal I.

4.4 Finite Type Morphisms

Let A be a noetherian commutative ring. A commutative A-algebra $A \to B$ is of *finite type* if B is isomorphic to quotient of a polynomial ring over A in finitely many variables:

$$
B \cong A[X_1, \ldots, X_m]/I.
$$

This globalizes in the evident manner: Let X be a noetherian scheme. A morphism $f : Y \to X$ is of *finite type* if X and Y admit finite affine covers $X = \cup_i U_i = \operatorname{Spec} A_i$, $Y = \cup_i V_i = \operatorname{Spec} B_i$ with $f(V_i) \subset U_i$ and $f^* : A_i \to B_i$ making B_i a finite-type A_i-algebra for each i.

In case $X = \operatorname{Spec} A$ for some noetherian ring A, we say that Y is of finite type over A. We let $\mathbf{Sch}_A$ denote the full subcategory of all schemes with objects the A-scheme of finite type which are separated over $\operatorname{Spec} A$.

Clearly the property of $f : Y \to X$ being of finite type is preserved under fiber product with an arbitrary morphism $Z \to X$ (with Z noetherian). By the Hilbert basis theorem, if $f : Y \to X$ is a finite type morphism, then Y is noetherian (X is assumed noetherian as part of the definition).

4.5 Proper, Finite and Quasi-Finite Morphisms

In topology, a proper morphism is one for which the inverse image of a compact set is compact. As above, the lack of good separation for the Zariski topology means one needs to use a somewhat different notion.

Definition 4.8 *A morphism $f : Y \to X$ is closed if for each closed subset C of $|Y|$, $f(C)$ is closed in $|X|$. A morphism $f : Y \to X$ is* proper *if*

1. *f is separated.*
2. *f is universally closed: for each morphism $Z \to X$, the projection $Z \times_X Y \to Z$ is closed.*

It is easy to see that the properness of a morphism $f : Y \to X$ is a local property on X.

Example 4.9 *A closed immersion $i : Z \to X$ is proper. Indeed, we may assume that X is affine, $X = \operatorname{Spec} A$, and that $Z = \operatorname{Spec} A/I$ for some ideal I. Then Z is separated over $\operatorname{Spec} \mathbb{Z}$, hence separated over X; if we have a morphism $Y = \operatorname{Spec} B \to X$ corresponding to a ring homomorphism $\phi : A \to B$, then $Y \times_X Z \to Y$ is the closed subscheme defined by $\phi(I)$. Thus i is universally closed.*

Remark 4.10 *Since the closed subsets in the Zariski topology are essentially loci defined by polynomial equations, the condition that a morphism $f : Y \to X$ is proper implies the "principle of elimination theory": Let C be a "closed algebraic locus" in $Z \times_X Y$. Then the set of $z \in |Z|$ for which there exist a $\tilde{z} \in C \subset Z \times_X Y$ is a closed subset. We will see below in the section on projective spaces how to construct examples of proper morphisms.*

Definition 4.11 *Let $f : Y \to X$ be a morphism of noetherian schemes. f is called* affine *if for each affine open subscheme $U \subset X$, $f^{-1}(U)$ is an affine open subscheme of Y. An affine morphism is called* finite *if for $U = \operatorname{Spec} A \subset X$ with $f^{-1}(U) = \operatorname{Spec} B$, the homomorphism $f^* : A \to B$ makes B into a finitely generated A-module.*

Definition 4.12 *Let $f : Y \to X$ be a morphism of noetherian schemes. f is called* quasi-finite *if for each $x \in |X|$, $|f|^{-1}(x)$ is a finite set.*

The property of a morphism $f : Y \to X$ being proper, affine, finite or quasi-finite is preserved under fiber product with an arbitrary morphism $Z \to X$, and the composition of two proper (resp. affine, finite, quasi-finite) morphisms is proper (resp. affine, finite, quasi-finite). Also

Proposition 4.13 *Let $X \xrightarrow{f} Y \xrightarrow{g} Z$ be morphisms of noetherian schemes. If gf is proper (resp. finite, quasi-finite) then f is proper (resp. finite, quasi-finite).*

A much deeper result is

Proposition 4.14 *Let $f : Y \to X$ be a finite type morphism. Then f is finite if and only if f is proper and quasi-finite.*

4.6 Flat Morphisms

The condition of flatness comes from homological algebra, but has geometric content as well. Recall that for a commutative ring A and an A-module M, the operation of tensor product $\otimes_A M$ is *right exact*:

$N' \to N \to N'' \to 0$ exact

$$\implies N' \otimes_A M \to N \otimes_A M \to N'' \otimes_A M \to 0 \text{ exact.}$$

M is a *flat* A-module if $\otimes_A M$ is exact, i.e., left-exact:

$$0 \to N' \to N \text{ exact } \implies 0 \to N' \otimes_A M \to N \otimes_A M \text{ exact.}$$

A ring homomorphism $A \to B$ is called flat if B is flat as an A-module.

Definition 4.15 *Let $f : Y \to X$ be a morphism of schemes. f is* flat *if for each $y \in Y$, the homomorphism $f_y^* : \mathcal{O}_{X,f(y)} \to \mathcal{O}_{Y,y}$ is flat.*

It follows from the "cancellation formula" $(N \otimes_A B) \otimes_B C \cong N \otimes_A C$ that a composition of flat morphisms is flat. Once we discuss dimension, we'll see that a flat morphism is locally equi-dimensional (see section 5.4).

4.7 Valuative Criteria

In the study of metric spaces, sequences and limits play a central role. In algebraic geometry, this is replaced by using the spectrum of *discrete valuation rings*.

Recall that a noetherian local domain $(\mathcal{O}, \mathfrak{m})$ is a discrete valuation ring (DVR for short) if $\mathfrak{m}$ is a principal ideal, $\mathfrak{m} = (t)$, $t \neq 0$. Let $\mathcal{O}$ have quotient field F and residue field k. It is not hard to see that $\operatorname{Spec} \mathcal{O}$ consists of two points: the generic point $\eta : \operatorname{Spec} F \to \operatorname{Spec} \mathcal{O}$ and the single closed point $\operatorname{Spec} k \to \operatorname{Spec} \mathcal{O}$. In terms of our sequence/limit analogy, $\operatorname{Spec} F$ is the sequence and $\operatorname{Spec} k$ is the limit. This should motivate the follow result which characterizes separated and proper morphisms.

Proposition 4.16 *Let $f : Y \to X$ be a morphism of finite type.*

1. *f is separated if and only if, for each DVR $\mathcal{O}$ and each commutative diagram*

$$\begin{array}{ccc} \operatorname{Spec} F & \longrightarrow & Y \\ \eta \downarrow & & \downarrow f \\ \operatorname{Spec} \mathcal{O} & \longrightarrow & X \end{array}$$

(where F is the quotient field of $\mathcal{O}$), there exists at most one lifting $\operatorname{Spec} \mathcal{O} \to Y$.

2. f is proper if and only if, for each DVR $\mathcal{O}$ and each commutative diagram

$$\begin{array}{ccc} \operatorname{Spec} F & \longrightarrow & Y \\ \eta \downarrow & & \downarrow f \\ \operatorname{Spec} \mathcal{O} & \longrightarrow & X \end{array}$$

(where F is the quotient field of $\mathcal{O}$), there exists a unique lifting $\operatorname{Spec} \mathcal{O} \to Y$.

5 The Category $\mathbf{Sch}_k$

In this section we fix a field k. The schemes of most interest to us in many applications are the separated k-schemes of finite type, i.e., the objects in the category $\mathbf{Sch}_k$. In this section we examine a number of concepts which one can describe quite concretely for such schemes.

5.1 R-Valued Points

The use of R-valued points allows one to recover the classical notion of "solutions of a system of equations" within the theory of schemes.

Definition 5.1 *Let X be a scheme, R a ring. The set of R-valued points $X(R)$ is by definition the Hom-set $\operatorname{Hom}_{\mathbf{Sch}}(\operatorname{Spec} R, X)$. If we fix a base-ring A, and X is a scheme over A and R is an A-algebra, we set*

$$X_A(R) := \operatorname{Hom}_{\mathbf{Sch}_A}(\operatorname{Spec} R, X).$$

We often leave off the subscript A if the context makes the meaning clear.

Example 5.2 *Let $X = \operatorname{Spec} k[X_1, \ldots, X_n]/(f_1, \ldots, f_r)$, and let F/k be an extension field of k. Then $X_k(F)$ is the set of maps*

$$\operatorname{Spec} F \to \operatorname{Spec} k[X_1, \ldots, X_n]/(f_1, \ldots, f_r)$$

over k, i.e., the set of k-algebra homomorphisms

$$\psi : k[X_1, \ldots, X_n]/(f_1, \ldots, f_r) \to F.$$

Clearly ψ is determined by the values $\psi(X_i)$, $i = 1, \ldots, n$; conversely, given elements $x_1, \ldots, x_n \in F$, we have the unique k-algebra homomorphism

$$\tilde{\psi} : k[X_1 \ldots, X_n] \to F$$

sending X_i to x_i. As $\tilde{\psi}(f_j) = f_j(x_1, \ldots, x_n)$, $\tilde{\psi}$ descends to an F-valued point ψ of X if and only if $f_j(x_1, \ldots, x_n) = 0$ for all j. Thus, we have identified the F-valued points of X with the set of solutions in F of the polynomial equations $f_1 = \ldots, f_r = 0$. This example explains the connection of the machinery of schemes with the basic problem of understanding the solutions of polynomial equations.

As a special case, take $X = \mathbb{A}_k^n$. Then $X_k(F) = F^n$ for all F.

5.2 Group-Schemes and Bundles

Just as in topology, we have the notion of a locally trivial bundle $E \to B$ with base B, fiber F and group G. The group G is an *algebraic group-scheme over* k, which is just a group-object in $\mathbf{Sch}_k$. Concretely, we have a multiplication $\mu : G \times_k G \to G$, inverse $\iota : G \to G$ and unit $e : \operatorname{Spec} k \to G$, satisfying the usual identities, interpreted as identities of morphisms.

Example 5.3 *Let $M_n = \operatorname{Spec} k[\{X_{ij} \mid 1 \leq i,j \leq n\}] \cong \mathbb{A}_k^{n^2}$. The formula for matrix multiplication*

$$\mu^*(X_{ij}) = \sum_k X_{ik} \otimes X_{kj}$$

defines the ring homomorphism

$$\mu^* : k[\ldots X_{ij} \ldots] \to k[\ldots X_{ij} \ldots] \otimes_k k[\ldots X_{ij} \ldots],$$

hence the morphism

$$\mu : M_n \times_k M_n \to M_n.$$

Let GL_n *be the open subset $(M_n)_{\det}$, where $\det$ is the determinant of the $n \times n$ matrix (X_{ij}), i.e.*

$$\operatorname{GL}_n := \operatorname{Spec} k[\ldots X_{ij} \ldots, \frac{1}{\det}].$$

Then μ restricts to

$$\mu : \operatorname{GL}_n \times_k \operatorname{GL}_n \to \operatorname{GL}_n,$$

and the usual formula for matrix inverse defines the inverse morphism $\iota : \operatorname{GL}_n \to \operatorname{GL}_n$. The unit is given by $e^*(X_{ij}) = \delta_{ij}$, giving us the group-scheme GL_n over k.

If G is an algebraic group-scheme over k, and F a finite type k-scheme, an action of G on F is just a morphism $\rho : G \times_k F \to F$, satisfying the usual associativity and unit conditions, again as identities of morphisms in $\mathbf{Sch}_k$.

Now we can just mimic the usual definition of a fiber-bundle with fiber F and group G: $p : E \to B$ is required to have local trivializations,

$$B = \cup_i U_i,$$

with the U_i open subschemes, and there are isomorphisms over U_i,

$$\psi_i : p^{-1}(U_i) \to U_i \times_k F.$$

In addition, for each i, j, there is a morphism $g_{ij} : U_i \cap U_j \to G$ such that the isomorphism

$$\psi_i \circ \psi_j^{-1} : (U_i \cap U_j) \times_k F$$

is given by the composition

$$(U_i \cap U_j) \times_k F \xrightarrow{(p_1, g_{ij} \circ p_1, p_2)} (U_i \cap U_j) \times_k G \times_k F \xrightarrow{\mathrm{id} \times \rho} (U_i \cap U_j) \times_k F.$$

An isomorphism of bundles $f : (E \to B) \to (E' \to B)$ is given by a B-morphism $f : E \to E'$ such that, with respect to a common local trivialization, f is locally of the form

$$U \times_k F \xrightarrow{(p_1, g \circ p_1, p_2)} U \times_k \times_k GF \xrightarrow{\mathrm{id} \times \rho} U \times_k F$$

for some morphism $g : U \to G$.

For example, using $G = \mathrm{GL}_n$, $F = \mathbb{A}_k^n$, $\rho : \mathrm{GL}_n \times_k \mathbb{A}_k^n \to \mathbb{A}_k^n$ the map with

$$\rho^*(Y_i) = \sum_j X_{ij} \otimes Y_j,$$

we have the notion of an *algebraic vector bundle* of rank n.

5.3 Dimension

Let A be a commutative ring. We have already discussed the Krull dimension of A: the maximal length n of a chain

$$\mathfrak{p}_0 \subset \mathfrak{p}_1 \subset \ldots \subset \mathfrak{p}_n$$

of distinct prime ideals in A. In this section, we describe other notions of dimension and how they relate to Krull dimension.

Let F be a finitely generated field extension of k. A set of elements of F are *transcendentally independent* if they satisfy no non-trivial polynomial identity with coefficients in k. A *transcendence basis* of F over k is a transcendentally independent set $\{x_\alpha \in F\}$ of elements of F such that F is algebraic over the subfield $k(\{x_\alpha\})$ generated by the x_α, i.e., each element $x \in F$ satisfies some non-trivial polynomial identity with coefficients in $k(\{x_\alpha\})$. The same argument that shows a vector space admits a basis and has a well-defined dimension shows that each F admits a transcendence basis over k and that each two transcendence bases of F over k have the same cardinality; the cardinality of a transcendence basis is called the *transcendence dimension* of F over k, $\mathrm{tr.\,dim}_k F$. Clearly if F is finitely generated over k, then $\mathrm{tr.\,dim}_k F$ is finite. In particular, if X is an integral k-scheme of finite type over k, then the function field $k(X)$ has finite transcendence dimension over k.

Definition 5.4 *Let X be an irreducible separated k-scheme of finite type. The dimension of X over k is defined by*

$$\dim_k X := \mathrm{tr.\,dim}_k k(X_{\mathrm{red}}).$$

In general, if X is a separated k-scheme of finite type with reduced irreducible components $X_1, \ldots, X_n$, we write

$$\dim_k X \leq d$$

if $\dim_k X_i \leq d$ for all i. We say that X is equi-dimensional over k *of dimension d if $\dim_k X_i = d$ for all i.*

Remark 5.5 *We can make the notion of dimension have a local character as follows: Let X be a separated finite type k-scheme and $x \in |X|$ a point. We say $\dim_k(X, x) \leq d$ if there is some neighborhood U of x in X with $\dim_k U \leq d$. We similarly say that X is equi-dimensional over k of dimension d at x if there is a U with $\dim_k U = d$. We say X is locally equi-dimensional over k if X is equi-dimensional over k at x for each $x \in |X|$.*

If X is locally equi-dimensional over k, then the local dimension function $\dim_k(X, x)$ is constant on connected components of $|X|$. Thus, if $W \subset X$ is an integral closed subscheme, then the local dimension function $\dim_k(X, w)$ is constant over W. We set $\mathrm{codim}_X W = \dim_k(X, w) - \dim_k W$. If w is the generic point of W and $\mathrm{codim}_X W = d$, we call w a codimension d point of X.

We thus have two possible definitions of the dimension of $\mathrm{Spec}\,A$ for A a domain which is a finitely generated k-algebra, $\dim_k \mathrm{Spec}\,A$ and the Krull dimension of A. Fortunately, these are the same:

Theorem 5.6 (Krull) *Let A be a domain which is a finitely generated k-algebra. Then $\dim_k \mathrm{Spec}\,A$ equals the Krull dimension of A.*

This result follows from the *principal ideal theorem* of Krull :

Theorem 5.7 (Krull) *Let A be a domain which is a finitely generated k-algebra, let f be a non-zero element of A and $\mathfrak{p} \supset (f)$ a minimal prime ideal containing (f). Then $\dim_k \mathrm{Spec}\,A = \dim_k \mathrm{Spec}\,A/\mathfrak{p} + 1$.*

the *Hilbert Nullstellensatz*:

Theorem 5.8 *Let A be a finitely generated k-algebra which is a field. Then $k \to A$ is a finite field extension.*

and induction.

5.4 Flatness and Dimension

There is a relation of flatness to dimension: Let $f : Y \to X$ be a finite type morphism of noetherian schemes. Then for each $x \in |X|$, $f^{-1}(x)$ is a scheme of finite type over the field $k(x)$, so one can ask if $f^{-1}(x)$ is equi-dimensional over $k(x)$ and if so, what is the dimension. If X and Y are irreducible and f is flat, then there is an integer $d \geq 0$ such that for each $x \in |X|$, either $f^{-1}(x)$ is empty or $f^{-1}(x)$ is an equi-dimensional $k(x)$-scheme of dimension d over $k(x)$. If $f : Y \to X$ is a flat morphism in $\mathbf{Sch}_k$, with X and Y equi-dimensional over k, then each non-empty fiber $f^{-1}(x)$ is equi-dimensional over $k(x)$, and

$$\dim_k Y = \dim_k X + \dim_{k(x)} f^{-1}(x).$$

Remark 5.9 *For a morphism $f : Y \to X$ in $\mathbf{Sch}_k$ with X and Y irreducible, the fiber dimension satisfies*

$$\dim_{k(x)} f^{-1}(x) \geq \dim_k Y - \dim_k X$$

if $f^{-1}(x) \neq \emptyset$. In addition, the fiber dimension is upper semi-continuous: if y is in the closure of x and $f^{-1}(y) \neq \emptyset$ then

$$\dim_{k(x)} f^{-1}(x) \geq d \Longrightarrow \dim_{k(x)} f^{-1}(y) \geq d.$$

5.5 Smooth Morphisms and étale Morphisms

The analog in algebraic geometry of a differentiable manifold of dimension n is a finite type k-scheme X which is *smooth* of dimension n over $\mathrm{Spec}\, k$. The cotangent bundle of a manifold is replaced by the sheaf of *relative Kähler differentials* for the structure morphism $f : X \to \mathrm{Spec}\, k$. For smooth k-schemes X and Y, the local algebraic theory is just like that for manifolds: a k-morphism $f : Y \to X$ is smooth if and only if f is a submersion. An *étale* morphism is the analog of a local isomorphism in the differentiable category; in particular, a proper étale morphism is the analog of a covering space. In this section, we give an introduction to this circle of ideas and constructions.

Let $\phi : A \to B$ be a ring homomorphism, and M a B-module. Recall that a *derivation* of B into M over A is an A-module homomorphism $\partial : B \to M$ satisfying the Leibniz rule

$$\partial(bb') = b\partial(b') + b'\partial(b).$$

Note that the condition of A-linearity is equivalent to $\partial(\phi(a)) = 0$ for $a \in A$.

The module of *Kähler differentials*, $\Omega_{B/A}$, is a B-module equipped with a universal derivation over A, $d : B \to \Omega_{B/A}$, i.e., for each derivation $\partial : B \to M$ as above, there is a unique B-module homomorphism $\psi : \Omega_{B/A} \to M$ with $\partial(b) = \psi(db)$. It is easy to construct $\Omega_{B/A}$: take the quotient of the free B-module on symbols db, $b \in B$, by the B-submodule generated by elements of the form

$$d(\phi(a)) \text{ for } a \in A$$
$$d(b + b') - db - db' \text{ for } b, b' \in B$$
$$d(bb') - b db' - b' db \text{ for } b, b' \in B.$$

Let $\bar{B} = B/I$ for some ideal I. We have the *fundamental exact sequence*:

$$I/I^2 \to \Omega_{B/A} \otimes_B \bar{B} \to \Omega_{\bar{B}/A} \to 0, \tag{5}$$

where the first map is induced by the map $f \mapsto df$.

Example 5.10 *Let k be a field, $B := k[X_1, \ldots, X_n]$ the polynomial ring over k. Then $\Omega_{B/k}$ is the free B-module on $dX_1, \ldots, dX_n$. If $B = k[X_1, \ldots, X_n]/I$ then the fundamental sequence shows that $\Omega_{B/k}$ is the quotient of $\oplus_{i=1}^n B \cdot dX_i$ by the submodule generated by $df = \sum_i \frac{\partial f}{\partial X_i} dX_i$ for $f \in I$.*

Definition 5.11 *Let $f : Y \to X$ be a morphism of finite type. f is smooth if*

1. *f is separated.*
2. *f is flat*
3. *Each non-empty fiber $f^{-1}(x)$ is locally equi-dimensional over $k(x)$.*
4. *Let y be in $|Y|$, let $x = f(y)$, let $d_y = \dim_{k(x)}(f^{-1}(x), y)$ and let $B_y = \mathcal{O}_{f^{-1}(x),y}$. Then $\Omega_{B_y/k(x)}$ is a free B_y-module of rank d_y.*

The map f is étale if f is smooth and $d_y = 0$ for all y.

Remarks 5.12 *(1) If X and Y are integral schemes, or if X and Y are in $\mathbf{Sch}_k$ and are both locally equi-dimensional over k, then the flatness of f implies the condition (3).*
(2) Smooth (resp. étale) morphisms are stable under base-change: if $f : Y \to X$ is smooth (resp. étale) and $Z \to X$ is an arbitrary morphism, then the projection $Z \times_X Y \to Z$ is smooth (resp. étale). There is a converse: A morphism $g : Z \to X$ is faithfully flat if g is flat and $|g| : |Z| \to |X|$ is surjective. A morphism $f : Y \to X$ is smooth (resp. étale) if and only if the projection $Z \times_X Y \to Z$ is smooth (resp. étale) for some faithfully flat $Z \to X$.

Examples 5.13 *(1) Let $k \to L$ be a finite extension of fields. Then $\mathrm{Spec}\, L \to \mathrm{Spec}\, k$ is smooth if and only if $\mathrm{Spec}\, L \to \mathrm{Spec}\, k$ is étale if and only if $k \to L$ is separable.*

(2) *Let x be a point of a scheme X, $m_{X,x} \subset \mathcal{O}_{X,x}$ the maximal ideal. The completion $\hat{\mathcal{O}}_{X,x}$ of $\mathcal{O}_{X,x}$ with respect to $m_{X,x}$ is the limit*

$$\hat{\mathcal{O}}_{X,x} := \varprojlim_{n} \mathcal{O}_{X,x}/m_{X,x}^{n}.$$

For example, if x is the point $(X_1, \ldots, X_n)$ in $X := \operatorname{Spec} k[X_1, \ldots, X_n]$, then $\hat{\mathcal{O}}_{X,x}$ is the ring of formal power series $k[[X_1, \ldots, X_n]]$.

Suppose k is algebraically closed and $f : Y \to X$ is a morphism of finite type k-schemes. Then f is étale if and only if for each closed point $y \in Y$, the map $f_y^* : \mathcal{O}_{X,f(y)} \to \mathcal{O}_{Y,y}$ induces an isomorphism on the completions $\hat{f}_y^* : \hat{\mathcal{O}}_{X,f(y)} \to \hat{\mathcal{O}}_{Y,y}$. In particular, if $X = \mathbb{A}_k^n$, then each $\hat{\mathcal{O}}_{Y,y}$ is isomorphic to a formal power series ring $k[[X_1, \ldots, X_n]]$.

Definition 5.14 *X in $\mathbf{Sch}_k$ is called a* smooth *k-scheme if the structure morphism $X \to \operatorname{Spec} k$ is smooth. We let $\mathbf{Sm}/k$ denote the full subcategory of $\mathbf{Sch}_k$ consisting of the smooth k-schemes. If X is in $\mathbf{Sch}_k$, we call a point $x \in |X|$ a* smooth point *if there is exists an open neighborhood U of x in X which is smooth over $\operatorname{Spec} k$.*

Let $\mathcal{O}$ be a noetherian local ring with maximal ideal $\mathfrak{m}$. Recall that a sequence of elements $t_1, \ldots, t_r$ in $\mathfrak{m}$ is a *regular sequence* if for each $i = 1, \ldots, r$ the image $\bar{t}_i$ in $\mathcal{O}/(t_1, \ldots, t_{i-1})$ is not a zero-divisor. It turns out that if $t_1, \ldots, t_r$ is a regular sequence, then so is each reordering of the sequence.

The local ring $\mathcal{O}$ is called *regular* if the maximal ideal is generated by a regular sequence. If $\mathcal{O} = \mathcal{O}_{X,x}$ for some point x on a scheme X, a choice $(t_1, \ldots, t_n)$ of a regular sequence generating $m_{X,x}$ is called a *system of local parameters* for X at x.

From the standpoint of homological algebra, the regular local rings are characterized by the theorem of Auslander-Buchsbaum as those for which the residue field $\mathcal{O}/m$ admits a finite free resolution, or equivalently, those for which *every* finitely generated $\mathcal{O}$-module admits a finite free resolution. The relation with smooth points is

Proposition 5.15 *Take $X \in \mathbf{Sch}_k$ and $x \in |X|$. If x is a smooth point, then $\mathcal{O}_{X,x}$ is a regular local ring, and the maximal ideal is generated by a regular sequence $(t_1, \ldots, t_n)$ with $n = \operatorname{codim}_X x$. Conversely, if k is perfect (k has characteristic 0 or k has characteristic $p > 0$ and $k^p = k$), and $\mathcal{O}_{X,x}$ is a regular local ring for some $x \in X$, then x is a smooth point of X.*

Similarly, one can view a system of parameters as follows:

Proposition 5.16 *Let X be in $\mathbf{Sch}_k$, $x \in X$ a closed point. Suppose X is equi-dimensional over k at x, and let $n = \dim_k(X, x)$. Take $t_1, \ldots, t_n$ in $m_{X,x}$*

and let U be an open neighborhood of x in X with $t_i \in \mathcal{O}_X(U)$ for each i, giving the morphism

$$f := (t_1, \ldots, t_n) : U \to \mathbb{A}_k^n$$

with $f(x) = 0$.

1. *If f is étale at x, then x is a smooth point of X and $t_1, \ldots, t_n$ is a system of parameters for $\mathcal{O}_{X,x}$*
2. *If k is perfect, then the following are equivalent:*
   a) *f is étale at x.*
   b) *x is a smooth point of X and $t_1, \ldots, t_n$ is a system of parameters for $\mathcal{O}_{X,x}$.*
   c) *The completion $\hat{\mathcal{O}}_{X,x}$ is a power-series ring in $t_1, \ldots, t_n$ over $k(x)$:*

$$\hat{\mathcal{O}}_{X,x} = k(x)[[t_1, \ldots, t_n]].$$

Another important property of smooth points on $X \in \mathbf{Sch}_k$ is that the set of smooth points forms an open subset of $|X|$. In particular $X \in \mathbf{Sch}_k$ is smooth over k if and only if each closed point of X is a smooth point. The closed subset of non-smooth (singular) points of X is denoted X_{sing}.

5.6 The Jacobian Criterion

The definition of a smooth point on an affine k-scheme $X \subset \mathbb{A}_k^n$ can be given in terms of the familiar criterion from differential topology. Suppose that X is defined by an ideal $I \subset k[X_1, \ldots, X_n]$, $I = (f_1, \ldots, f_r)$. Let x be a closed point of X. For $g \in k[X_1, \ldots, X_n]$, we have the "value" $g(x) \in k(x)$, where $g(x)$ is just the image of g under the residue homomorphism $k[X_1, \ldots, X_n] \to k(x)$. In particular, we can evaluate the Jacobian matrix of the f_i's at x forming the matrix

$$\mathrm{Jac}(x) := \left(\frac{\partial f_i}{\partial X_j} \right)(x) \in M_{n \times n}(k(x)).$$

Proposition 5.17 *Let $X := \mathrm{Spec}\, k[X_1, \ldots, X_n]/(f_1, \ldots, f_r)$. Then $x \in X$ is a smooth point if and only if X is equi-dimensional over k at x and*

$$\mathrm{rank}(Jac(x)) = n - \dim_k(X, x).$$

The proof follows by considering the fundamental exact sequence (5).

6 Projective Schemes and Morphisms

Classical algebraic geometry deals with algebraic subsets of projective space. In this section, we describe the modern machinery for constructing closed subschemes of projective spaces and, more generally, projective morphisms.

6.1 The Functor Proj

The functor Spec is the basic operation going from rings to schemes. We describe a related operation Proj from graded rings to schemes.

Recall that a (non-negatively) graded ring is a ring R whose underlying additive group is a direct sum, $R = \oplus_{n=0}^{\infty} R_n$, such that the multiplication respects the grading:

$$R_n \cdot R_m \subset R_{n+m}.$$

We assume all our rings are commutative, so R is automatically an R_0-algebra.

An element of R_n is said to be homogeneous of degree n. An ideal $I \subset R$ is called *homogeneous* if $I = \sum_{n=0}^{\infty} I \cap R_n$; we often write I_n for $R_n \cap I$. Note that I is homogeneous if and only if the following condition holds:

If f is in I, and we write $f = \sum_n f_n$ with $f_n \in R_n$, then each f_n is also in I.

(6)

If R is a graded ring and $I \subset R$ a homogeneous ideal, then R/I is also graded, $R/I = \oplus_{n=0}^{\infty} R_n/I_n$.

Example 6.1 *Fix a ring A, and let $R = A[X_0, \ldots, X_m]$, where we give each X_i degree 1. Then R has the structure of a graded ring, with $R_0 = A$, and R_n the free A-module with basis the monomials $X_0^{d_0} \cdot \ldots \cdot X_m^{d_m}$ of total degree $n = \sum_i d_i$. Unless we make explicit mention to the contrary, we will always use this structure of a graded ring on $A[X_0, \ldots, X_m]$.*

Fix a noetherian ring A. We consider graded A-algebras $R = \oplus_{n=0}^{\infty} R_n$ such that

1. $R_0 = A \cdot 1$, i.e. R_0 is generated as an A-module by 1,
2. R is generated as an A-algebra by R_1, and R_1 is finitely generated as an A-module.

Equivalently, if R_1 is generated over A by elements $r_0, \ldots, r_m$, then sending X_i to r_i exhibits R as a (graded) quotient of $A[X_0, \ldots, X_m]$. Letting I be the kernel of the surjection $A[X_0, \ldots, X_m] \to R$, we see that I is a graded ideal, so there are homogeneous polynomials $f_1, \ldots, f_r \in A[X_0, \ldots, X_m]$ with $I = (f_1, \ldots, f_r)$.

Let R be a graded A-algebra satisfying (1) and (2). We let $R^+ \subset R$ be the ideal $\oplus_{n \geq 1} R_n$. Define the set $\operatorname{Proj} R$ to be the set of homogeneous prime ideals $\mathfrak{p} \subset R$ such that $\mathfrak{p}$ does not contain R^+. For a homogeneous ideal I, we let

$$V_h(I) = \{\mathfrak{p} \in \operatorname{Proj} R \mid \mathfrak{p} \supset I\},$$

The operation V_h has properties analogous to the properties for V listed in (1), so we can define a topology on $\operatorname{Proj} R$ for which the closed subsets are exactly those of the form $V_h(I)$, for I a homogeneous ideal..

We now define a sheaf of rings on $\operatorname{Proj} R$. For this, we use a homogeneous version of localization. Let S be a subset of R. If S is a multiplicatively closed subset of R, containing 1, we define $S_h^{-1}R$ to be the ring of fractions f/s with $s \in S_n := S \cap R_n$, $f \in R_n$, $n = 0, 1, \ldots$, modulo the usual relation

$$f/s = f'/s' \text{ if } s''(s'f - sf') = 0 \text{ for some } s'' \in S_{n''}.$$

Note that $S_h^{-1}R$ is just a commutative ring, we have lost the grading.

Let $Y = \operatorname{Proj} R$. For $f \in R_n$, we have the open subset $Y_f := Y \setminus V_h((f))$. Let $S(f) = \{1, f, f^2, \ldots\}$ and set $\mathcal{O}_Y(Y_f) := S(f)_h^{-1}R$. This forms the "partially defined" sheaf on the principal open subsets Y_f. If $U = Y \setminus V_h(I)$ is now an arbitrary open subset of $\operatorname{Proj} R$, we set

$$\mathcal{O}_Y(U) := \ker\left(\prod_{\substack{f \in I \\ f \text{ homogeneous}}} \mathcal{O}_Y(Y_f) \to \prod_{\substack{f,g \in I \\ f,g \text{ homogeneous}}} \mathcal{O}_Y(Y_{fg})\right)$$

where the map is the difference of the two restriction maps. Just as for affine scheme, this defines a sheaf of rings $\mathcal{O}_Y$ on Y with the desired value $\mathcal{O}_Y(Y_f) = S(f)_h^{-1}R$ on the principal open subsets Y_f.

Lemma 6.2 Let f be in R_n. Then $(Y_f, (\mathcal{O}_Y)_{|Y_f}) \cong \operatorname{Spec} S(f)_h^{-1}R$ as ringed spaces.

Proof (sketch of proof). Let $Z = \operatorname{Spec} S(f)_h^{-1}R$. Let $J \subset R$ be a homogeneous ideal. Form the ideal $J_f \subset S(f)_h^{-1}R$ as the set of elements g/f^m, $g \in J_{nm}$. Conversely, let $I \subset S(f)_h^{-1}R$ be an ideal. Let $I^h \subset R$ be the set of elements of the form g, with $g \in R_{nm}$ and $g/f^m \in I$. Then I^h is a homogeneous ideal in R.

One checks the relations:

$$(I^h)_f = I; \quad (J_f)^h \supset J.$$

In addition, the operations $I \mapsto I^h$, $J \mapsto J_f$ send prime ideals to prime ideals, and if $\mathfrak{q} \subset R$ is a homogeneous prime, $\mathfrak{q} \not\supset (f)$, then $(\mathfrak{q}_f)^h = \mathfrak{q}$. Thus, we have the bijection between Y_f and Z, which one easily sees is a homeomorphism. Under this homeomorphism, the open subset Y_{fg}, $g \in R_{nm}$, corresponds to the open subset Z_{g/f^m}. Similarly, the isomorphism

$$\mathcal{O}_Z(Z_{g/f^m}) = S(g/f^m)^{-1}(S(f)_h^{-1}R) \cong S(fg)_h^{-1}R = \mathcal{O}_Y(Y_{fg})$$

shows that we can extend our homeomorphism to an isomorphism of ringed spaces $Y_f \cong Z$.

Now take $\mathfrak{p} \in Y = \operatorname{Proj} R$, and take some element $f \in R_1 \setminus \mathfrak{p}_1$. Then $\mathfrak{p}$ is in Y_f; by the lemma above, this gives us an affine open neighborhood of $\mathfrak{p}$. Thus $\operatorname{Proj} R$ is a scheme.

Sending $a \in A$ to $a/f^0 \in \mathcal{O}_Y(Y_f)$ gives the ring homomorphism $p^* : A \to \mathcal{O}_Y(Y)$, and hence the *structure morphism* $p : \operatorname{Proj} R \to \operatorname{Spec} A$.

Example 6.3 *We take the most basic example, namely*

$$R = k[X_0, \ldots, X_n],$$

k a field. The scheme $\operatorname{Proj} R$ is then the projective *n-space over k, $\mathbb{P}_k^n \to$ $\operatorname{Spec} k$. We have the affine open cover $\mathbb{P}_k^n = \cup_{i=0}^n U_i$, where $U_i = (\operatorname{Proj} R)_{X_i} = \operatorname{Spec} k[X_0/X_i, \ldots, X_n/X_i]$. As $k[X_0/X_i, \ldots, X_n/X_i]$ is clearly a polynomial ring over k in variables X_j/X_i, $j \neq i$, we have the isomorphisms $U_i \cong \mathbb{A}_k^n$. The change of coordinates in passing from U_i to U_j is just*

$$(X_m/X_j) = (X_m/X_i)(X_i/X_j),$$

which is the same as the standard patching data for the complex or real projective spaces.

We have a similar description of the F-valued points of $\mathbb{P}_k^n$, for F/k an extension field. Indeed, if $f : \operatorname{Spec} F \to \mathbb{P}_k^n$ is a morphism over $\operatorname{Spec} k$, then, as $|\operatorname{Spec} F|$ is a single point, f must factor through some $U_i \subset \mathbb{P}_k^n$. Thus, we have the F-valued point of $\operatorname{Spec} k[X_0/X_i, \ldots, X_n/X_i]$, i.e., a homomorphism $\psi : k[X_0/X_i, \ldots, X_n/X_i] \to F$, which is given by the values $\psi(X_m/X_i) = x_m^{(i)}$ for $m \neq i$, $\psi(X_i/X_i) = x_i^{(i)} = 1$. If we make a different choice of affine open U_j, we have the point $(x_0^{(j)}, \ldots, x_n^{(j)})$ with $x_m^{(j)} = x_m^{(i)}/x_j^{(i)}$ for $m = 0, \ldots, n$. Thus, we have the familiar description of $\mathbb{P}_k^n(F)$ as

$$\mathbb{P}_k^n(F) = \{x = (x_0, \ldots, x_n) \in F^{n+1} \setminus \{0\}\}/x \sim \lambda \cdot x, \ \lambda \in F \setminus \{0\}.$$

We denote the equivalence class of a point $(x_0, \ldots, x_n)$ by $(x_0 : \ldots : x_n)$.

It is not hard to see that a (graded) surjection of graded A-algebras $R \to \bar{R}$ gives rise to a closed immersion $\operatorname{Proj} \bar{R} \to \operatorname{Proj} R$, and this in turn identifies the collection of closed subschemes of $\operatorname{Proj} R$ with the collection of homogeneous ideals $J \subset R$, where we identify two such ideals J and J' if the localizations J_x, J'_x agree for all $x \in R_1$. For example, if we have homogeneous polynomials $f_1, \ldots, f_r \in k[X_0, \ldots, X_n]$, these generate a homogeneous ideal $J = (f_1, \ldots, f_r)$, and give us the closed subscheme $Y := \operatorname{Proj} k[X_0, \ldots, X_n]/J$ of $\mathbb{P}_k^n$. The F-valued points of Y are exactly the F-valued points $(x_0 : \ldots : x_n)$ with $f_j(x_0, \ldots, x_n) = 0$ for all j.

Remark 6.4 *The equivalence relation on homogeneous ideas described above using localization is just*

$$\operatorname{Proj} k[X_0, \ldots, X_n]/J = \operatorname{Proj} k[X_0, \ldots, X_n]/J'$$
$$\text{as closed subschemes of } \mathbb{P}_k^n$$
$$\Longleftrightarrow J_m = J'_m \text{ for } m >> 0.$$

More generally, if $R = \oplus_{n \geq 0} R_n$ is a graded A algebra satisfying our conditions (1) and (2), choosing A-module generators $r_0, \ldots, r_n$ for R_1 defines the surjection of graded A-algebras $\pi : A[X_0, \ldots, X_n] \to R$ and thus identifies $\operatorname{Proj} R$ with the closed subscheme of $\mathbb{P}_A^n$ defined by the homogeneous ideal $\ker \pi$.

6.2 Properness

The main utility of Proj is that it gives a direct means of constructing proper morphisms without going to the trouble of explicitly gluing affine schemes.

Proposition 6.5 *Let R be a graded A-algebra satisfying (1) and (2) above. Then the structure morphism $p : \operatorname{Proj} R \to \operatorname{Spec} A$ is a proper morphism of finite type..*

Proof. If $f_0, \ldots, f_m$ generate R_1 over A, then the finite affine open cover

$$\operatorname{Proj} R = \cup_{i=0}^{m}(\operatorname{Proj} R)_{f_i}$$

exhibits p as a morphism of finite type. To check that p is proper, we use the valuative criterion of proposition 4.16.

Sending X_i to f_i defines a graded surjection $k[X_0, \ldots, X_m] \to R$, that is, a closed immersion $i : \operatorname{Proj} R \to \mathbb{P}_A^m$ making

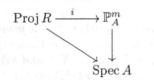

commute. As a closed immersion is proper, it suffices to show that $\mathbb{P}_A^m \to \operatorname{Spec} A$ is proper.

So, let $\mathcal{O}$ be a DVR with quotient field F and maximal ideal (t), and suppose we have a commutative diagram

$$\begin{array}{ccc} \operatorname{Spec} F & \xrightarrow{f} & \mathbb{P}_A^m \\ \eta \downarrow & & \downarrow p \\ \operatorname{Spec} \mathcal{O} & \xrightarrow{g} & \operatorname{Spec} A \end{array}$$

Replacing A with $\mathcal{O}$, we may assume that $g = \operatorname{id}$.

One can extend our characterization of the F-valued points of $\mathbb{P}_{\mathcal{O}}^m$ to the $\mathcal{O}$-valued points as follows: The $\mathcal{O}$-valued points of $\mathbb{P}_{\mathcal{O}}^m$ are $n + 1$-tuples $(r_0, \ldots, r_m)$ of elements of $\mathcal{O}$ with not all r_i in (t), modulo multiplication by units in $\mathcal{O}$.

The F-valued point f of $\mathbb{P}_{\mathcal{O}}^m$ consists of an $m+1$-tuple $(f_0, \ldots, f_m)$, $f_i \in F$, with not all $f_i = 0$, modulo scalar multiplication by $F^\times$. Write each f_i as

$$f_i = u_i t^{n_i}$$

where (t) is the maximal ideal of $\mathcal{O}$, the u_i are units and the n_i integers. Letting n be the minimum of the n_i, $(u_0 t^{n_0 - n}, \ldots, u_m t^{n_m - n})$ gives the same F-valued point as f and all the coordinates are in $\mathcal{O}$, not all in (t), giving us a lifting $\operatorname{Spec} \mathcal{O} \to \mathbb{P}_{\mathcal{O}}^m$. Our characterization of $\mathbb{P}_{\mathcal{O}}^m(\mathcal{O})$ also proves uniqueness of the lifting.

6.3 Projective and Quasi-Projective Morphisms

Definition 6.6 *Let X be a noetherian k-scheme. A morphism $f : Y \to X$ is called* projective *if f admits a factorization $f = p \circ i$, where $p : \mathbb{P}_k^n \times_k X \to X$ is the projection and $i : Y \to: \mathbb{P}_k^n \times_k X$ is a closed immersion. A morphism $f : Y \to X$ is called* quasi-projective *if f admits a factorization $f = \bar{f} \circ j$, with $j : Y \to \bar{Y}$ an open immersion and $\bar{f} : \bar{Y} \to X$ a projective morphism.*

A k-scheme X is called a projective *(resp.* quasi-projective*) k-scheme if the structure morphism $p : X \to \operatorname{Spec} k$ is projective (resp. quasi-projective).*

Proposition 6.7 *A projective morphism is a proper morphism of finite type.*

Proof. Since a closed immersion is a proper morphism of finite type, it suffices to prove the case of the projection $\mathbb{P}_k^n \times_k X \to X$, which follows from proposition 6.5 and the fact that the property of a morphism being proper and of finite type is preserved by arbitrary base-change.

Remark 6.8 (The Segre embedding) *The projective property is preserved by taking products. In fact, consider the tensor product $k[X_0, \ldots, X_n] \otimes_k k[Y_0, \ldots, Y_M]$ where we give $k[X_0, \ldots, X_n]$ and $k[Y_0, \ldots, Y_M]$ the grading by degree. Consider the polynomial ring $k[\{Z_{ij} | 0 \le i \le N, 0 \le j \le M\}]$; we have the homomorphism*

$$\phi_{N,M}^* : k[Z_{ij}] \to k[X_0, \ldots, X_n] \otimes_k k[Y_0, \ldots, Y_M]$$

sending Z_{ij} to $X_i \otimes Y_j$. It is not hard to see that the kernel of $\phi_{N,M}^*$ is a homogeneous ideal, and that $\phi_{N,M}^*$ defines a closed immersion

$$\phi_{N,M} : \mathbb{P}_k^N \times_k \mathbb{P}_k^M \to \mathbb{P}^{(N+1)(M+1)-1},$$

called the Segre embedding*. Restricting to closed subschemes $X \subset \mathbb{P}_k^N$ and $Y \subset \mathbb{P}_k^M$, we see that, if X and Y are projective k-schemes, so is $X \times_k Y$ a projective k-scheme.*

Remark 6.9 (Elimination theory) *Using the principle of elimination theory for a proper morphism (remark 4.10) and proposition 6.7, we have the classical statement of elimination theory: Setting $\mathbb{P}_k^N := \operatorname{Proj} k[X_0, \ldots, X_N]$, $\mathbb{P}_k^M := \operatorname{Proj} k[Y_0, \ldots, Y_M]$, suppose that $C \subset \mathbb{P}_k^N \times_k \mathbb{P}_k^M$ is defined by polynomial equations $f_\ell(X, Y)$ in the X_i and Y_j which are separately homogeneous in the X_i and in the Y_j:*

$$f_\ell(\lambda X, \mu Y) = \lambda^d \mu^e f_\ell(X, Y).$$

Then $p_1(C) \subset \mathbb{P}_k^N$ is defined by homogeneous equations in the X_i.

Classically, the explicit elimination of the Y_j was performed by using resultants.

Example 6.10 *Fix positive integers M and d. The k-vector space of homogeneous polynomials in $Y_0, \ldots, Y_M$ of degree d has basis the monomials of degree d, and is thus a k-vector space of dimension $\binom{M+d}{d}$. For an index $I := (i_0, \ldots, i_M)$ with $i_j \geq 0$, $\sum_j i_j = d$, we have the corresponding monomial Y^I; let $\{X_I\}$ be a set of variables indexed by such I.*

Setting $N := \binom{M+d}{d} - 1$, we thus have the universal hypersurface of degree d, H, as the closed subscheme of $\mathbb{P}_k^N \times_k \mathbb{P}_k^M$ defined by the bi-homogeneous polynomial

$$F(X; Y) := \sum_I X_I Y^I.$$

Clearly the fiber H_a of H over a point $a := (\ldots : a_I : \ldots)$ of $\mathbb{P}_k^N$ is the degree d hypersurface H_a of $\mathbb{P}_k^M$ defined by $F(a; Y)$.

Let H^{sing} be the closed subscheme of $\mathbb{P}_k^N \times_k \mathbb{P}_k^M$ defined by the bi-homogeneous ideal generated by $F(X; Y)$ and all the partial derivatives $\partial F(X; Y)/\partial Y_j$, $j = 0, \ldots, M$. By the Jacobian criterion for smoothness, $H^{sing} \cap H_a$ is the set of singular (i.e., non-smooth) points of H_a. Thus, the projection

$$disc_{d,M} := p_1(H^{sing})$$

is the set of points $a \in \mathbb{P}_k^N$ such that H_a has a singular point.

By the principle of elimination theory, $disc_{d,M}$ is a closed subset of $\mathbb{P}_k^N$, hence defined by a homogeneous ideal. In fact, for $d \geq 2$, $disc_{d,M}$ has pure codimension one and is thus defined by a single homogeneous equation. For the case $M = 1$, this is (up to a scalar) the classical discriminant of polynomials of degree d.

6.4 Globalization

One can use the operation Proj to define proper morphisms over non-affine schemes as well. One simply replaces graded A-algebras with graded sheaves of $\mathcal{O}_X$-algebras. If $\mathcal{R} = \oplus_{n \geq 0} \mathcal{R}_n$ is a sheaf of graded $\mathcal{O}_X$-algebras for some noetherian scheme X, then $\mathcal{R}(U)$ is a graded $\mathcal{O}_X(U)$-algebra for all open subschemes U of X. We require that X admits an affine open cover $X = \cup_i U_i := \operatorname{Spec} A_i$ such that the graded A_i-algebra $\mathcal{R}(U_i)$ satisfies our conditions (1) and (2) for each i. The A_i-schemes $\operatorname{Proj}_{A_i} \mathcal{R}(U_i)$ then patch together to give the X-scheme $p : \operatorname{Proj}_{\mathcal{O}_X} \mathcal{R} \to X$. p is clearly a proper morphism, as properness is a local property on the base scheme.

One can show that, in case X is a quasi-projective scheme over a field k, then $\operatorname{Proj}_{\mathcal{O}_X} \mathcal{R} \to X$ is actually a projective morphism, i.e., there exists a closed immersion $\operatorname{Proj}_{\mathcal{O}_X} \mathcal{R} \to \mathbb{P}_k^N \times_k X$ for some $N >> 0$. Thus, we are not getting any new schemes over X by this added generality, however, this does make some useful constructions more natural.

Example 6.11 *Let $\mathcal{E}$ be a rank $n + 1$ locally free sheaf of $\mathcal{O}_X$-modules on a noetherian scheme X. Form the sheaf of symmetric algebras $\mathcal{R} := \mathrm{Sym}^*_{\mathcal{O}_X}(\mathcal{E})$. Take an affine open cover $X = \cup_i U_i$ trivializing $\mathcal{E}$. If $U_i = \mathrm{Spec}\, A_i$, a choice of an isomorphism $\mathrm{res}_{U_i}\mathcal{E} \cong \mathcal{O}^{n+1}_{U_i}$ gives an isomorphism*

$$\mathcal{R}(U_i) \cong A_i[X_0, \ldots, X_n].$$

On $U_i \cap U_j$ the isomorphism $\mathrm{res}_{U_i \cap U_j}\mathrm{res}_{U_i}\mathcal{E} \cong res_{U_i \cap U_j}\mathrm{res}_{U_j}\mathcal{E}$ yields a change of coordinates in the variables X_l, given by an invertible matrix $g_{ij} \in \mathrm{GL}_{n+1}(\mathcal{O}_X(U_i \cap U_j))$. This data gives us a rank $n + 1$ vector bundle $E \to X$ with sheaf of sections isomorphic to $\mathcal{E}$, and one has the isomorphism of X-schemes

$$\mathrm{Proj}_{\mathcal{O}_X}\mathrm{Sym}^*_{\mathcal{O}_X}(\mathcal{E}) \cong \mathbb{P}(E^\vee),$$

where $E^\vee$ is the dual bundle and $\mathbb{P}(E^\vee) \to X$ is the fiber bundle with fiber the projective space $\mathbb{P}(E^\vee_x)$ over a point $x \in X$,

$$\mathbb{P}(E^\vee_x) := E^\vee_x \setminus \{0\}/v \sim \lambda v; \ \lambda \in k(x)^*.$$

We write $\mathbb{P}(\mathcal{E})$ for $\mathrm{Proj}_{\mathcal{O}_X}\mathrm{Sym}^*_{\mathcal{O}_X}(\mathcal{E})$.

6.5 Blowing Up a Subscheme

A very different type of Proj is the blow-up of a subscheme $Z \subset X$ with X in $\mathbf{Sch}_k$. Let $\mathcal{I}_Z$ be the ideal sheaf defining Z. The X-scheme $\pi : \mathrm{Bl}_Z X \to X$ is defined as

$$\mathrm{Bl}_Z X := \mathrm{Proj}_{\mathcal{O}_X}(\oplus_{n \geq 0}\mathcal{I}^n_Z),$$

where we give $\mathcal{I}^*_Z := \oplus_{n \geq 0}\mathcal{I}^n_Z$ the structure of a graded $\mathcal{O}_X$-algebra by using the multiplication maps $\mathcal{I}^n_Z \times \mathcal{I}^m_Z \to \mathcal{I}^{n+m}_Z$. As we have seen, π is a proper morphism, and is projective if for example X is quasi-projective over k.

To analyze the morphism π, let $U = X \setminus Z$. The restriction of $\mathcal{I}_Z$ to U is $\mathcal{O}_U$, so the restriction of $\mathcal{I}^*_Z$ to U is $\oplus_{n \geq 0}\mathcal{O}_U$ with the evident multiplication. This is just the graded $\mathcal{O}_U$-algebra $\mathcal{O}_U[X_0]$ ($\deg X_0 = 1$); using the evident correspondence between graded ideals in $A[X_0]$ and ideals in A, for a commutative ring A, we see that

$$\pi^{-1}(U) = \mathrm{Proj}_{\mathcal{O}_U}(\mathcal{O}_U[X_0]) = U,$$

with π the identity map. Over Z, something completely different happens: $Z \times_X \mathrm{Bl}(X, Z)$ is just $\mathrm{Proj}_{\mathcal{O}_Z}(\mathcal{I}^*_Z \otimes_{\mathcal{O}_X} \mathcal{O}_Z)$. Since $\mathcal{O}_Z = \mathcal{O}_X/\mathcal{I}_Z$, we find

$$\mathcal{I}^n_Z \otimes_{\mathcal{O}_X} \mathcal{O}_X/\mathcal{I}_Z = \mathcal{I}^n_Z/\mathcal{I}^{n+1}_Z,$$

hence

$$\pi^{-1}(Z) = \mathrm{Proj}_{\mathcal{O}_Z}(\oplus_{n \geq 0}\mathcal{I}^n_Z/\mathcal{I}^{n+1}_Z).$$

The sheaf of $\mathcal{O}_Z$-modules $\mathcal{I}_Z/\mathcal{I}^2_Z$ is called the *conormal sheaf* of Z in X. If Z is locally defined by a regular sequence of length d then $\mathcal{I}_Z/\mathcal{I}^2_Z$ is a locally free sheaf of $\mathcal{O}_Z$-modules (of rank d); the dual of the corresponding vector bundle on Z is the *normal bundle* of Z in X, $N_{Z/X}$. For example, if both Z and X are smooth over k, and $d = \text{codim}_X Z$, then Z is locally defined by a regular sequence of length d and $N_{Z/X}$ is the usual normal bundle from differential topology.

Assuming that Z is locally defined by a regular sequence of length d, we have the isomorphism of graded $\mathcal{O}_Z$-algebras

$$\oplus_{n \geq 0} \mathcal{I}_Z^n / \mathcal{I}_Z^{n+1} \cong \text{Sym}^*(\mathcal{I}_Z / \mathcal{I}_Z^2).$$

Thus $\pi^{-1}(Z)$ is a $\mathbb{P}^{d-1}$-bundle over Z, in fact

$$\pi^{-1}(Z) = \mathbb{P}(N_{Z/X}).$$

Thus, we have "blown-up" Z in X by replacing Z with the projective space bundle of normal lines to Z.

The simplest example is the blow-up of the origin $0 \in \mathbb{A}_k^m$. This yields

$$\text{Bl}_Z X$$
$$= \text{Proj}_{k[X_1,\ldots,X_m]}(\oplus_{n \geq 0}(X_1,\ldots,X_m)^n)$$
$$= \text{Proj}_{k[X_1,\ldots,X_n]}(k[X_1,\ldots,X_m,Y_1,\ldots,Y_m]/(\ldots X_i Y_j - X_j Y_i \ldots)),$$

where we give the Y_i's degree 1. The fiber over 0 is thus

$$\text{Proj}_k(k[Y_1,\ldots,Y_m]) = \mathbb{P}_k^{m-1},$$

which we identify with the projective space of lines in $\mathbb{A}_k^m$ through 0.

At this point, we should mention the fundamental result of Hironaka on resolution of singularities:

Theorem 6.12 (Hironaka [8]) *Let k be an algebraically closed field of characteristic zero, X a reduced finite type k-scheme. Then there is a sequence of morphisms of reduced finite type k-schemes*

$$X_N \to X_{N-1} \to \ldots \to X_1 \to X_0 = X$$

and reduced closed subschemes $Z_n \subset X_{n \, \text{sing}}$, $n = 0,\ldots,N-1$ such that

1. *each Z_n is smooth over k.*
2. *$X_{n+1} \to X_n$ is the morphism $\text{Bl}_{Z_n} X_n \to X_n$.*
3. *X_N is smooth over k.*

The hypothesis that k be algebraically closed was later removed, but the result in characteristic $p > 0$ and in mixed characteristic is still an important open problem.

II

Sheaves for a Grothendieck Topology

A presheaf on a topological space T is a contravariant functor on the category $\mathrm{Op}(T)$ of open subsets of T; a sheaf S is just a presheaf satisfying a patching condition and a locality condition, namely, the exactness of

$$S(U) \xrightarrow{\prod_\alpha \mathrm{res}_{U_\alpha,U}} \prod_\alpha S(U_\alpha) \underset{\prod_{\alpha,\beta} \mathrm{res}_{U_\alpha \cap U_\beta, U_\beta}}{\overset{\prod_{\alpha,\beta} \mathrm{res}_{U_\alpha \cap U_\beta, U_\alpha}}{\rightrightarrows}} \prod_{\alpha.\beta} S(U_\alpha \cap U_\beta)$$

for each open cover $U = \cup_\alpha U_\alpha$ of each open subset $U \subset T$. Grothendieck showed how to generalize this construction to a small category equipped with an extra structure, which is given by *covering families* or *covering sieves*. In this part, we recall Grothendieck's theory and discuss its main points.

The applications to motivic homotopy theory that appear in this book will mainly use the example of the Nisnevich topology on the category of smooth schemes of finite type over a field. For a more thorough discussion of the main topic of this chapter, and for more details on properties of étale cohomology, we direct the interested reader to [1], [2], [3] and [11].

Notation. Sets is the category of sets, **Ab** the category of abelian groups. For k a field, we let $\mathbf{Sch}_k$ denote the category of separated schemes of finite type over k, and $\mathbf{Sm}/k$ the full subcategory of smooth schemes over k.

7 Limits

Before discussing presheaves and sheaves, we need some basic results on limits.

7.1 Definitions

Let I be a small category, $\mathcal{A}$ a category and $F : I \to \mathcal{A}$ a functor. Form the category $\varinjlim F$ with objects

$$(X, \{f_i : F(i) \to X \mid i \in I\})$$

with X in $\mathcal{A}$ and with the condition that, for each morphism $g : i \to i'$ in I, we have $f_{i'} \circ F(g) = f_i$. A morphism

$$\phi : (X, \{f_i : F(i) \to X\}) \to (X', \{f_i' : F(i) \to X'\})$$

is a morphism $\phi : X \to X'$ in $\mathcal{A}$ with $f_i' = \phi \circ f_i$ for all i. Dually, we have the category $\varprojlim F$ with objects

$$(X, \{f_i : X \to F(i) \mid i \in I\})$$

such that for each morphism $g : i \to i'$ in I, we have $f_{i'} = F(g) \circ f_i$; a morphism

$$\phi : (X, \{f_i : X \to F(i)\}) \to (X', \{f_i' : X' \to F(i)\})$$

is a morphism $\phi : X \to X'$ in $\mathcal{A}$ with $f_i' \circ \phi = f_i$ for all i.

Definition 7.1 *Let I be a small category, $\mathcal{A}$ a category and $F : I \to \mathcal{A}$ a functor. The* inductive limit $\varinjlim F$ *is an initial object in $\varinjlim F$, and the* projective *limit* $\varprojlim F$ *is a final object in $\varprojlim F$.*

Example 7.2 *For $\mathcal{A} = \mathbf{Sets}$ we have explicit expressions for $\varinjlim F$ and $\varprojlim F$: $\varinjlim F$ is the quotient of the disjoint union $\coprod_{i \in I} F(i)$ by the relation*

$$x_i \in F(i) \sim F(g)(x_i) \in F(i')$$

for $g : i \to i'$ in I, and $\varprojlim F$ is the subset of $\prod_{i \in I} F(i)$ consisting of elements $\prod_i x_i$ with $F(g)(x_i) = x_{i'}$ for $g : i \to i'$ in I.

For $\mathcal{A} = \mathbf{Ab}$, replacing disjoint union with direct sum in the above yields $\varinjlim F$; the projective limit $\varprojlim F$ is defined by exactly the same formula as for $\mathbf{Sets}$.

One can express the universal property of $\varinjlim F$ and $\varprojlim F$ in terms of limits of sets by the formulas (for fixed Z in $\mathcal{A}$):

$$\mathrm{Hom}_{\mathcal{A}}(\varinjlim F, Z) \cong \varprojlim \mathrm{Hom}_{\mathcal{A}}(F(-), Z) \qquad (7)$$

$$\mathrm{Hom}_{\mathcal{A}}(Z, \varprojlim F) \cong \varprojlim \mathrm{Hom}_{\mathcal{A}}(Z, F(-)) \qquad (8)$$

Here $\mathrm{Hom}_{\mathcal{A}}(F(-), Z) : I^{\mathrm{op}} \to \mathbf{Sets}$ and $\mathrm{Hom}_{\mathcal{A}}(Z, F(-)) : I \to \mathbf{Sets}$ are the functors $i \mapsto \mathrm{Hom}_{\mathcal{A}}(F(i), Z)$ and $i \mapsto \mathrm{Hom}_{\mathcal{A}}(Z, F(i))$, respectively, and the isomorphisms are induced by the structure morphisms $F(i) \to \varinjlim F$, $\varprojlim F \to F(i)$. In fact

Proposition 7.3 *The identities* (7) *and* (8) *characterize* $\lim_{\rightarrow} F$ *and* $\lim_{\leftarrow} F$, *respectively, i.e.,* $\lim_{\rightarrow} F$ *together with the maps* $F(i) \rightarrow \lim_{\rightarrow} F$, $i \in I$, *represents the functor*

$$Z \mapsto \lim_{\leftarrow} \mathrm{Hom}_{\mathcal{A}}(F(-), Z)$$

and $\lim_{\leftarrow} F$ *together with the maps* $\lim_{\leftarrow} F \rightarrow F(i)$, $i \in I$, *represents the functor*

$$Z \mapsto \lim_{\leftarrow} \mathrm{Hom}_{\mathcal{A}}(Z, F(-)).$$

7.2 Functoriality of Limits

Let $F : I \rightarrow \mathcal{A}$ and $G : I \rightarrow \mathcal{A}$ be functors. A natural transformation of functors $\theta : F \rightarrow G$ yields functors $\theta^* : \lim_{\rightarrow} G \rightarrow \lim_{\rightarrow} F$ and $\theta_* : \lim_{\leftarrow} F \rightarrow \lim_{\leftarrow} G$, thus a morphism on the initial, resp. final, objects

$$\theta_* : \lim_{\rightarrow} F \rightarrow \lim_{\rightarrow} G; \ \ \theta_* : \lim_{\leftarrow} F \rightarrow \lim_{\leftarrow} G$$

assuming these exist. These satisfy $(\theta \circ \theta')_* = \theta_* \circ \theta'_*$.

Similarly, a functor $f : J \rightarrow I$ induces $f_* : \lim_{\rightarrow} F \circ f \rightarrow \lim_{\rightarrow} F$ and $f^* :$ $\lim_{\leftarrow} F \rightarrow \lim_{\leftarrow} F \circ f$, with $(ff')_* = f_* \circ f'_*$ and $(ff')^* = f'^* \circ f^*$.

7.3 Representability and Exactness

A functor $G : \mathcal{C} \rightarrow \mathcal{D}$ is called *left exact* if G commutes with finite projective limits, *right exact* if G commutes with finite inductive limits and *exact* if both left and right exact. From our formulas for $\lim_{\rightarrow} F$ and $\lim_{\leftarrow} F$, we have:

Proposition 7.4 *For* $\mathcal{A} = \mathbf{Sets}, \mathbf{Ab}$:

1. $\lim_{\rightarrow} F$ *and* $\lim_{\leftarrow} F$ *both exist for arbitrary functors* $F : I \rightarrow \mathcal{A}$.
2. $F \mapsto \lim_{\underset{I}{\rightarrow}} F$ *is right exact*
3. $F \mapsto \lim_{\underset{I}{\leftarrow}} F$ *is left exact.*

In general, $F \mapsto \lim_{\underset{I}{\rightarrow}} F$ is not left-exact, however, under special conditions on I, this is the case. The usual conditions are:

L1. Given $i \rightarrow j$, $i \rightarrow j'$ in I, there exists a commutative square

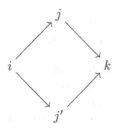

L2. Given morphisms $i \overset{f}{\underset{g}{\rightrightarrows}} j$ there is a morphism $h : j \to k$ with $hf = hg$.

L3. Give i, i', there are morphisms $i \to j$, $i' \to j$.

A category I satisfying $L1 - L3$ is called a *filtering* category. The main result on filtered inductive limits is

Proposition 7.5 *Let I be a small filtering category, $\mathcal{A} = $ Sets or $\mathcal{A} = $ Ab. Then $F \mapsto \varinjlim_I F$ is exact.*

For a proof, see [3].

7.4 Cofinality

It is often useful to replace an index category I with a subcategory $\epsilon : J \to I$. In general, this changes inductive and projective limits, but there is a criterion which ensures that $\varinjlim_I F = \varinjlim_J F \circ \epsilon$.

Definition 7.6 *A subcategory J of a small category I is called* cofinal *if*

1. *Given $i \in I$, there is a morphism $i \to j$ with $j \in J$.*
2. *J is a full subcategory of I.*

Here the main result is:

Lemma 7.7 *Suppose that $\epsilon : J \to I$ is a cofinal subcategory of a small category I. Then for $F : I \to$ Sets or $F : I \to$ Ab, the map*

$$\epsilon_* : \varinjlim_J F \circ \epsilon \to \varinjlim_I F$$

is an isomorphism.

8 Presheaves

Fix a small category $\mathcal{C}$ and a category $\mathcal{A}$. A *presheaf P* on $\mathcal{C}$ with values in a category $\mathcal{A}$ is a functor

$$P : \mathcal{C}^{\mathrm{op}} \to \mathcal{A}.$$

Morphisms of presheaves are natural transformations of functors. This defines the category of $\mathcal{A}$-valued presheaves on $\mathcal{C}$, $PreShv^{\mathcal{A}}(\mathcal{C})$.

Remark 8.1 *We require $\mathcal{C}$ to be small so that the collection of natural transformations $\vartheta : F \to G$, for presheaves F, G, form a set. It would suffice that $\mathcal{C}$ be essentially small (the collection of isomorphism classes of objects form a set). In practice, one often ignores the smallness condition on $\mathcal{C}$.*

8.1 Limits and Exactness

One easily sees that the existence, exactness and cofinality of limits in **Sets** or **Ab** (proposition 7.4, proposition 7.5, lemma 7.7) is inherited by the presheaf category:

Proposition 8.2 *For* $\mathcal{B} = PreShv^{\mathbf{Sets}}(\mathcal{C})$ *or* $\mathcal{B} = PreShv^{\mathbf{Ab}}(\mathcal{C})$:

1. $\varinjlim F$ *and* $\varprojlim F$ *both exist for arbitrary functors* $F : I \to \mathcal{B}$.
2. $F \mapsto \varinjlim F$ *is right exact*
3. $F \mapsto \varprojlim F$ *is left exact.*
4. *If I is filtering, then* $F \mapsto \varinjlim F$ *is exact*
5. *If* $\epsilon : J \to I$ *is cofinal, then* $\epsilon_* : \varinjlim_{J} F \circ \epsilon \to \varinjlim_{I} F$ *is an isomorphism.*

Indeed, since inductive and projective limits are functors, we have the formulas (for $F : I \to PreShv^{\mathcal{A}}(\mathcal{C})$)

$$(\varinjlim F)(X) = \varinjlim_{i \in I} F(i)(X)$$

$$(\varprojlim F)(X) = \varprojlim_{i \in I} F(i)(X).$$

8.2 Functoriality and Generators for Presheaves

Let $f : \mathcal{C} \to \mathcal{C}'$ be a functor of small categories. Composition with f defines the presheaf pull-back

$$f^p : PreShv^{\mathcal{A}}(\mathcal{C}') \to PreShv^{\mathcal{A}}(\mathcal{C}).$$

Lemma 8.3 *For* $\mathcal{A} = \mathbf{Sets}, \mathbf{Ab}$, f^p *has a left adjoint*

$$f_p : PreShv^{\mathcal{A}}(\mathcal{C}) \to PreShv^{\mathcal{A}}(\mathcal{C}').$$

f_p *is right-exact and* f^p *is exact.*

Proof. Recall that a functor $L : \mathcal{B} \to \mathcal{B}'$ is left-adjoint to a functor $R : \mathcal{B}' \to \mathcal{B}$ (equivalently, R is right adjoint to L) if there are isomorphisms

$$\theta_{X,Y} : \mathrm{Hom}_{\mathcal{B}}(X, R(Y)) \to \mathrm{Hom}_{\mathcal{B}'}(L(X), Y) \tag{9}$$

natural in X and Y. Taking $Y = L(X)$ and $X = R(Y)$ and applying the isomorphism to the respective identity elements, we find morphisms $\psi_X : X \to R \circ L(X)$ and $\phi_Y : L \circ R(Y) \to Y$ which yield the isomorphism $\theta_{X,Y}$ by $(f : X \to R(Y)) \mapsto \phi_Y \circ L(f)$ and its inverse by $(g : L(X) \to Y) \mapsto R(g) \circ \psi_X$. Conversely, given natural transformations $\psi : \mathrm{id} \to R \circ L$ and $\phi : L \circ R \to \mathrm{id}$, L and R form an adjoint pair if the compositions

$$R \xrightarrow{\psi \circ R} R \circ L \circ R \xrightarrow{R \circ \phi} R$$

$$L \xrightarrow{L \circ \psi} L \circ R \circ L \xrightarrow{\phi \circ L} L$$

are the identity natural transformations. It follows immediately from the existence of the natural isomorphism (9) that L is right-exact and R is left-exact.

To define the functor f_p, take $Y \in \mathcal{C}'$ and let I_Y be the category with objects pairs (X, ϕ), where $\phi : Y \to f(X)$ is a morphism in $\mathcal{C}'$. A morphism $g : (X, \phi) \to (X', \phi')$ is a morphism $g : X \to X'$ in $\mathcal{C}$ with $\phi' = f(g) \circ \phi$. I_Y is a small category.

For $F \in PreShv^{\mathcal{A}}(\mathcal{C})$, let $F_Y : I_Y^{\mathrm{op}} \to \mathcal{A}$ be the functor $F_Y(X, \phi) = F(X)$ and set

$$(f_p F)(Y) := \varinjlim F_Y.$$

A morphism $g : Y' \to Y$ in $\mathcal{C}'$ gives the functor $g^* : I_Y \to I_{Y'}$ by $g^*(X, \phi) = (X, \phi \circ g)$, and we have the identity $F_Y = F_{Y'} \circ g^*$. Thus, the functor

$$(g^*)_* : \varinjlim F_Y \to \varinjlim F_{Y'}$$

gives the morphism

$$(f_p F)(g) : (f_p F)(Y) \to (f_p F)(Y')$$

satisfying $(f_p F)(gg') = (f_p F)(g') \circ (f_p F)(g)$. We have therefore constructed a presheaf $f_p F$ on $\mathcal{C}'$.

To show that f_p is left-adjoint to f^p, we need to construct natural maps $\psi_F : F \to f^p f_p F$ and $\phi_G : f_p f^p G \to G$ for $F \in PreShv^{\mathcal{A}}(\mathcal{C})$ and $G \in PreShv^{\mathcal{A}}(\mathcal{C}')$. For this, note that $f^p f_p F(X) = (f_p F)(f(X))$. We have the object $(X, \mathrm{id}_{f(X)})$ in $I_{f(X)}$, giving the canonical map

$$F(X) = F_{f(X)}(X, \mathrm{id}_{f(X)}) \to \varinjlim F_{f(X)} = (f_p F)(f(X)),$$

which defines ψ_F. For $Y \in \mathcal{C}'$, $f_p f^p G(Y)$ is the limit over I_Y of the functor $(X, \phi) \mapsto G(f(X))$. For each $(X, \phi) \in I_Y$, we have the map $G(\phi) : G(f(X)) \to G(Y)$; the universal property of $\varinjlim$ thus yields the desired map $\phi_G(Y) : f_p f^p G(Y) \to G(Y)$. One checks the necessary compatibility $(R \circ \phi) \circ (\psi \circ R) = \mathrm{id}_R$, $(\phi \circ L) \circ (L \circ \psi) = \mathrm{id}_L$ without trouble.

The right-exactness of f_p follows from the adjoint property; the exactness of f^p follows from the fact that limits in the presheaf category are taken pointwise: $(\lim F)(X) = \lim F(X)$.

8.3 Generators for Presheaves

Recall that a *set of generators* for a category $\mathcal{C}$ is a set of objects $\mathcal{S}$ of $\mathcal{C}$ with the property that, if $f, g : A \to B$ are morphisms in $\mathcal{C}$ with $f \neq g$, then there exists an $X \in \mathcal{S}$ and a morphism $h : X \to A$ with $fh \neq gh$. If $\mathcal{S} = \{X\}$, X is called a generator for $\mathcal{C}$.

Example 8.4 Sets *has as generator the one-point set 0, and* **Ab** *has the generator* $\mathbb{Z}$. *If inductive limits exist in* $\mathcal{C}$ *and* $\mathcal{C}$ *has a set of generators* $\mathcal{S}$, *then* $\coprod_{X \in \mathcal{S}} X$ *is a generator for* $\mathcal{C}$.

Let $\mathcal{C}$ be a small category. We will use the functor f_p to construct generators for the presheaf categories $PreShv^{\mathcal{A}}(\mathcal{C})$, $\mathcal{A} =$ **Sets, Ab**.

Take X in $\mathcal{C}$, let $*$ denote the one-point category (with only the identity morphism) and $i_X : * \to \mathcal{C}$ the functor with $i_X(*) = X$. Since $PreShv^{\mathcal{A}}(*) = \mathcal{A}$, we have the functor

$$i_{Xp} : \textbf{Sets} \to PreShv^{\textbf{Sets}}(\mathcal{C}).$$

Let $\varXi_X = i_{Xp}(0)$. Similarly, we have $i_{Xp} : \textbf{Ab} \to PreShv^{\textbf{Ab}}(\mathcal{C})$; we set $\mathcal{Z}_X := i_{Xp}(\mathbb{Z})$. As $i_X^p(P) = P(X)$, the adjoint property of i_{Xp} and i_X^p gives

$$P(X) = \text{Hom}_{\textbf{Sets}}(0, i_X^p(P)) = \text{Hom}_{PreShv^{\textbf{Sets}}(\mathcal{C})}(\varXi_X, P)$$
$$P(X) = \text{Hom}_{\textbf{Ab}}(\mathbb{Z}, i_X^p(P)) = \text{Hom}_{PreShv^{\textbf{Ab}}(\mathcal{C})}(\mathcal{Z}_X, P)$$

for P in $PreShv^{\textbf{Sets}}(\mathcal{C})$, resp. P in $PreShv^{\textbf{Ab}}(\mathcal{C})$. Clearly this shows

Proposition 8.5 $\{\varXi_X \mid X \in \mathcal{C}\}$ *is a set of generators for* $PreShv^{\textbf{Sets}}(\mathcal{C})$ *and* $\{\mathcal{Z}_X \mid X \in \mathcal{C}\}$ *is a set of generators for* $PreShv^{\textbf{Ab}}(\mathcal{C})$.

8.4 $PreShv^{\textbf{Ab}}(\mathcal{C})$ as an Abelian Category

We begin this section by recalling some of the basic facts on abelian categories.

Let $\mathcal{A}$ be an additive category, i.e., the Hom-sets are given the structure of abelian groups such that, for each morphism $f : X \to Y$, and object Z of $\mathcal{A}$, the maps $f_* : \text{Hom}(Z, X) \to \text{Hom}(Z, Y)$ and $f^* : \text{Hom}(Y, Z) \to \text{Hom}(X, Z)$ are group homomorphisms. In addition, one requires that finite coproducts (direct sums) exist. One shows that this implies that finite products also exist, and products and coproducts agree. In particular, $\mathcal{A}$ has an initial and final object 0.

If $\mathcal{A}$ is an additive category, and $f : X \to Y$ a morphism, $i : \ker f \to X$ is a morphism which is universal for maps $g : Z \to X$ such that $fg = 0$, i.e. there exists a unique morphism $\phi : Z \to \ker f$ such that $i\phi = g$. Dually, $j : Y \to \text{coker} f$ is universal for morphisms $h : Y \to Z$ such that $hf = 0$, in that there is a unique morphism $\psi : \text{coker} f \to Z$ with $h = \psi j$. These morphisms, if they exist, are called the *categorical kernel* and *categorical cokernel*, respectively.

An *abelian category* is an additive category $\mathcal{A}$ such that each morphism $f : X \to Y$ admits a categorical kernel and cokernel, and the canonical map

$$\text{coker}(\ker f) \to \ker(\text{coker} f)$$

is an isomorphism. The object $\ker(\text{coker} f)$ is the *image* of f, denoted $\text{im} f$. The primary example of an abelian category is the category Mod_R of (left) modules over a ring R, for example **Ab** $:= \text{Mod}_{\mathbb{Z}}$.

A *complex* in an abelian category $\mathcal{A}$ is a sequence of morphisms

$$\ldots \xrightarrow{d^{n-2}} M^{n-1} \xrightarrow{d^{n-1}} M^n \xrightarrow{d^n} M^{n+1} \xrightarrow{d^{n+1}} \ldots$$

with $d^n \circ d^{n-1} = 0$ for all n. Under this condition, the map d^{n-1} factors as

$$M^{n-1} \to \operatorname{im} d^{n-1} \to \ker d^n \to M^n.$$

The *cohomology* of a complex (M^*, d^*) is defined by

$$H^n(M^*, d^*) := \operatorname{coker}(\operatorname{im} d^{n-1} \to \ker d^n),$$

i.e., the familiar quotient object $\ker d^n / \operatorname{im} d^{n-1}$. We call the complex *acyclic* if all cohomology objects H^n vanish. Viewing a complex as a sequence of maps, one also refers to an acyclic complex as an *exact sequence*.

An *injective* object in an abelian category $\mathcal{A}$ is an object I such that, for each monomorphism $i : N' \to N$, the map $i^* : \operatorname{Hom}_{\mathcal{A}}(N, I) \to \operatorname{Hom}_{\mathcal{A}}(N', I)$ is surjective. Dually a *projective* object in $\mathcal{A}$ is an object P such that, for each epimorphism $j : N \to N'$ the map $j_* : \operatorname{Hom}_{\mathcal{A}}(P, N) \to \operatorname{Hom}_{\mathcal{A}}(P, N')$ is surjective. We say that $\mathcal{A}$ *has enough injectives* if each object M of $\mathcal{A}$ admits a monomorphism $M \to I$ with I injective; $\mathcal{A}$ *has enough projectives* if each M admits an epimorphism $P \to M$ with P projective.

These conditions are useful, as we shall see later, for defining right and left derived functors of left- and right-exact functors. For now, we will only recall that a *right-resolution* of an object M in an abelian category $\mathcal{A}$ is an acyclic complex of the form

$$0 \to M \to I^0 \to \ldots \to I^n \to \ldots$$

and dually a *left-resolution* is an acyclic complex of the form

$$\ldots \to P^n \to \ldots \to P^0 \to M \to 0.$$

A right-resolution with all I^n injective is an *injective* resolution of M; a left-resolution with all P^n projective is a *projective* resolution of M. Clearly, if $\mathcal{A}$ has enough injectives, then each M in $\mathcal{A}$ admits an injective resolution, dually if $\mathcal{A}$ has enough projectives.

We now return to our discussion of the presheaf category. Let $\mathcal{C}$ be a small category, $\mathcal{A}$ an abelian category and $f : F \to G$ a map in $PreShv^{\mathcal{A}}(\mathcal{C})$. Since the categorical ker and coker are defined by universal properties, it is clear that $X \mapsto \ker(f(X) : F(X) \to G(X))$ and $X \mapsto \operatorname{coker}(f(X) : F(X) \to G(X))$ define $\mathcal{A}$-valued presheaves $\ker f$, $\operatorname{coker} f$ on $\mathcal{C}$ and that these are the respective categorical kernel and cokernel of f. Thus

Proposition 8.6 *Let $\mathcal{A}$ be an abelian category. Then $PreShv^{\mathcal{A}}(\mathcal{C})$ is an abelian category where, for $f : F \to G$ a morphism, $\ker f$, resp. $\operatorname{coker} f$ are the presheaves*

$$(\ker f)(X) = \ker f(X); \quad (\operatorname{coker} f)(X) = \operatorname{coker} f(X).$$

We recall the basic result of Grothendieck on the existence of enough injective objects:

Theorem 8.7 (Grothendieck [6]) *Let $\mathcal{A}$ be an abelian category. Suppose that*

1. *(small) inductive limits exist in $\mathcal{A}$.*
2. *if I is a small filtering category, then $(F : I \to \mathcal{A}) \mapsto \varinjlim F$ is exact.*
3. *$\mathcal{A}$ has a set of generators.*

Then $\mathcal{A}$ has enough injectives.

Remark 8.8 *The condition (1) is equivalent to the condition (AB3) in [6]; the conditions (1) and (2) together is equivalent to the condition (AB5) in [6].*

Proposition 8.9 *For a small category $\mathcal{C}$, the abelian category $PreShv^{\mathbf{Ab}}(\mathcal{C})$ has enough injectives.*

Proof. The conditions (1) and (2) of theorem 8.7 follow from proposition 8.2. Condition (3) is proposition 8.5.

9 Sheaves

We recall the definition of a Grothendieck pre-topology on a category $\mathcal{C}$, and the resulting category of sheaves. Unless explicitly mentioned to the contrary, we will assume that the value category $\mathcal{A}$ is either **Sets** or **Ab**; we leave it to the reader to make the necessary changes for more general value categories, such as simplicial sets, G-**Sets** for a group G or Mod_R for a ring R. See the remarks at the end of section 9.2 for a discussion of presheaves and sheaves of simplicial sets.

9.1 Grothendieck Pre-Topologies and Topologies

For a category $\mathcal{C}$ and object X in $\mathcal{C}$, we let $\mathcal{C}/X$ denote the category of morphisms $Y \to X$ in $\mathcal{C}$, i.e. the objects are morphisms $f : Y \to X$ and a morphism $g : (f : Y \to X) \to (f' : Y' \to X)$ is a morphism $g : Y \to Y'$ in $\mathcal{C}$ with $f = f' \circ g$.

Definition 9.1 *Let $\mathcal{C}$ be a category. A Grothendieck pre-topology τ on $\mathcal{C}$ is given by the following data: for each object $X \in \mathcal{C}$ there is a set $\mathrm{Cov}_\tau(X)$ of covering families of X, where a covering family of X is a set of morphisms $\{f_\alpha : U_\alpha \to X \mid \alpha \in A\}$ in $\mathcal{C}$. The sets $\mathrm{Cov}_\tau(X)$ should satisfy the following axioms:*

A1. $\{\mathrm{id}_X\}$ is in $\mathrm{Cov}_\tau(X)$ for each $X \in \mathcal{C}$.

A2. For $\{f_\alpha : U_\alpha \to X\} \in \mathrm{Cov}_\tau(X)$ and $g : Y \to X$ a morphism in $\mathcal{C}$, the fiber products $U_\alpha \times_X Y$ all exist and $\{p_2 : U_\alpha \times_X Y \to Y\}$ is in $\mathrm{Cov}_\tau(Y)$.

A3. If $\{f_\alpha : U_\alpha \to X\}$ is in $\mathrm{Cov}_\tau(X)$ and if $\{g_{\alpha\beta} : V_{\alpha\beta} \to U_\alpha\}$ is in $\mathrm{Cov}_\tau(U_\alpha)$ for each α, then $\{f_\alpha \circ g_{\alpha\beta} : V_{\alpha\beta} \to X\}$ is in $\mathrm{Cov}_\tau(X)$.

We will not need the notion of a Grothendieck topology in order to make the construction of primary interest for us, namely sheaves. Roughly speaking, giving a pre-topology is analogous to giving a basis of open subsets for a topology on a set; in particular each Grothendieck pre-topology generates a Grothendieck topology. For this reason, we will often omit the distinction between a pre-topology and a topology on a category. However, just for completeness, we briefly recall the definition of a Grothendieck topology.

Definition 9.2 *(1) Let $\mathcal{C}$ be a (small) category, X an object in $\mathcal{C}$. A* sieve on X *is a subfunctor of* $\mathrm{Hom}_\mathcal{C}(-, X)$, *i.e. a set $\mathcal{S}$ of objects of $\mathcal{C}/X$ such that, for each $f : Y \to X$ in $\mathcal{S}$ and each morphism $g : Z \to Y$ in $\mathcal{C}$, the composition $f \circ g : Z \to X$ is in $\mathcal{S}$.*

(2) Let $\mathcal{F} := \{f : Y_i \to X\}$ be a collection of morphisms in $\mathcal{C}$. The sieve generated by $\mathcal{F}$ *is the collection of morphisms $h : Y \to X$ which admit a factorization $h = f \circ g$ for some f in $\mathcal{F}$.*

(3) If $\mathcal{S}$ is a sieve over X and $f : Z \to X$ is a morphism in $\mathcal{C}$, the restriction $\mathcal{S}_Z$ *of $\mathcal{S}$ to Z is the sieve over Z generated by the maps $g : Y \to Z$ such that $f \circ g$ is in $\mathcal{S}$.*

Definition 9.3 *Let $\mathcal{C}$ be a (small) category. A* Grothendieck topology *on $\mathcal{C}$ consists in giving, for each object X of $\mathcal{C}$ a family $\mathcal{S}(X)$ of sieves over X, called* covering sieves, *such that*

a1. The sieve generated by id_X is a covering sieve of X.

a2. If $\mathcal{S}$ is a covering sieve of X and $f : Z \to X$ is a morphism in $\mathcal{C}$, then $\mathcal{S}_Z$ is a covering sieve of Z.

a3. Let $\mathcal{S}$ be a covering sieve of X. If T is a sieve over X such that, for each $f : Y \to X$ in $\mathcal{S}$, the restriction T_Y is a covering sieve of Y, then T is a covering sieve of X.

A category $\mathcal{C}$ together with a Grothendieck topology on $\mathcal{C}$ is called a site.

If we are given for each $X \in \mathcal{C}$ a family $\mathcal{S}_0(X)$ of sieves over X, there is clearly a minimal Grothendieck topology $X \mapsto \mathcal{S}(X)$ with $\mathcal{S}(X) \supset \mathcal{S}_0(X)$ for all X; this is the Grothendieck topology *generated by the $\mathcal{S}_0(X)$.*

If we are given a Grothendieck pre-topology τ on $\mathcal{C}$, we have, for each X in $\mathcal{C}$ and each covering family $\{f_\alpha : U_\alpha \to X\} \in \mathrm{Cov}_\tau(X)$, the sieve generated by the covering family. The resulting collection of sieves generates a Grothendieck topology on $\mathcal{C}$.

From now on, we will define all our Grothendieck topologies via covering families, and not distinguish between a Grothendieck pre-topology and a Grothendieck topology.

Examples 9.4 *(1) The "classical" example is the topology T on $\mathcal{C} = \mathrm{Op}(T)$, for T a topological space. Here $\mathrm{Op}(T)$ is the category of open subsets of T, with morphisms corresponding to inclusions of subsets. For $U \subset T$ open, $\{f_\alpha : U_\alpha \to U\}$ is in $\mathrm{Cov}_T(U)$ if $U = \cup_\alpha U_\alpha$. A somewhat less classical example is to let $\mathcal{C}$ be the category of topological spaces* **Top** *(this is not a small category, but let's ignore this). Setting $\mathrm{Cov}_{\mathbf{Top}}(U) = \mathrm{Cov}_U(U)$ gives a topology on* **Top**. *Note that for $U_\alpha \subset U$ open and $f : V \to U$ a continuous map, the fiber product $V \times_U U_\alpha$ is just the open subset $f^{-1}(U_\alpha)$ of V. In particular, for $\{f_\alpha : U_\alpha \to U\}$ in $\mathrm{Cov}(U)$, the fiber product $U_\alpha \times_U U_\beta$ is just the intersection $U_\alpha \cap U_\beta$.*

(2) We will be mostly interested in Grothendieck topologies on subcategories of **Sch**$_k$, *k a field. The first example is the category $\mathrm{Op}(|X|)$, $X \in$ **Sch**$_k$, with the covering families as in (1), where $|X|$ denotes the topological space underlying the scheme X. This gives the Zariski topology on X; we denote this site by X_{Zar}. We can extend this topology to all of* **Sch**$_k$, *as in example (1), giving the "big" Zariski site* **Sch**$_{k\mathrm{Zar}}$.

*(3) Take $X \in$ **Sch**$_k$. The étale site $X_{\text{ét}}$ has as underlying category the schemes over X, $f : U \to X$ such that f is an étale morphism of finite type. For such a U, a covering family $\{f_\alpha : U_\alpha \to U\}$ is a set of étale morphisms (of finite type) such that $\coprod_\alpha |U_\alpha| \to |U|$ is surjective. As in (2) one can use the same definition of covering families to define the big étale site* **Sch**$_{k\text{ét}}$.

(4) The Nisnevich *topology is between the étale and Zariski topologies: an étale covering family $\{f_\alpha : U_\alpha \to U\}$ is a Nisnevich covering family if for each field extension $L \supset k$, the induced map on the L-valued points*

$$\coprod_\alpha U_\alpha(L) \to U(L)$$

is surjective; this condition clearly implies that $\{f_\alpha : U_\alpha \to U\}$ is in $\mathrm{Cov}_{\text{ét}}(U)$. This defines the Nisnevich topology on X. We denote the associated site by X_{Nis}. The same definition of covering families defines the Nisnevich topology on **Sch**$_k$, *with associated site* **Sch**$_{k\mathrm{Nis}}$.

One can phrase the condition on the L-valued points somewhat differently: For each point $w \in U$, there is an α and a $w_\alpha \in U_\alpha$ with $f_\alpha(w_\alpha) = w$ and with $f_\alpha^* : k(w) \to k(w_\alpha)$ an isomorphism. This condition for a particular $w \in |U|$ with closure $W \subset U$ is equivalent to saying that for some α, the map $f_\alpha : U_\alpha \to U$ admits a section over some dense open subscheme W^0 of W.

(5) The h-topology *and variants (see [15, Chap. 2, §4]). We recall that a*

map of schemes $f : Y \to X$ *is a* topological epimorphism *if the map on the underlying spaces* $|f| : |Y| \to |X|$ *identifies* $|X|$ *with a quotient of* $|Y|$ *($|f|$ is surjective and* $U \subset |X|$ *is open if and only if* $|f|^{-1}(U)$ *is open in* $|Y|$*).* f *is a* universal topological epimorphism *if* $p_2 : Y \times_X Z \to Z$ *is a topological epimorphism for all* $Z \to X$.

The *covering families in the h-topology are finite sets* $\{f_i : U_i \to X\}$ *such that each* f_i *is of finite type and* $\coprod_i U_i \to X$ *is a universal topological epimorphism.*

The *covering families in the qfh-topology are the h-coverings* $\{f_i : U_i \to X\}$ *such that each* f_i *is quasi-finite.*

The *covering families in the cdh-topology are those generated (i.e., by iteratively applying the axioms A2 and A3) by*

a) *Nisnevich covering families*

b) *families of the form* $\{X' \amalg F \xrightarrow{p \amalg i} X\}$*, with* $p : X' \to X$ *proper,* $i : F \to X$ *a closed immersion and* $p : p^{-1}(X \setminus i(F)) \to X \setminus i(F)$ *an isomorphism.*

(6) *The* indiscrete *topology on a category* $\mathcal{C}$ *is the topology* ind *with*

$$\mathrm{Cov}_{ind}(X) = \{\{\mathrm{id}_X\}\}.$$

Remark 9.5 *Suppose we have a Grothendieck topology* τ *on a category* $\mathcal{C}$. *Let* $\mathcal{C}_0$ *be a full subcategory of* $\mathcal{C}$ *such that, if* X *is in* $\mathcal{C}_0$, $g : Y \to X$ *is a morphism is* $\mathcal{C}_0$ *and* $\{f_\alpha : U_\alpha \to X \mid \alpha \in A\}$ *is in* $\mathrm{Cov}_\tau(X)$*, then each* U_α *is in* $\mathcal{C}_0$ *and each fiber product* $Y \times_X U_\alpha$ *is in* $\mathcal{C}_0$. *Then we can restrict* τ *to* $\mathcal{C}_0$, *defining the Grothendieck topology* $\tau_{\mathcal{C}_0}$ *on* $\mathcal{C}_0$, *by setting*

$$\mathrm{Cov}_{\tau_{\mathcal{C}_0}}(X) = \mathrm{Cov}_\tau(X)$$

for X *in* $\mathcal{C}_0$.

As *an important example, let* $\mathbf{Sm}/k$ *denote the full subcategory of* $\mathbf{Sch}_k$ *consisting of the smooth k-schemes of finite type. Then the Zariski, étale and Nisnevich topologies on* $\mathbf{Sch}_k$ *restrict to Grothendieck topologies on* $\mathbf{Sm}/k$, *giving us the sites* $\mathbf{Sm}/k_{\mathrm{Zar}}$, $\mathbf{Sm}/k_{\text{ét}}$ *and* $\mathbf{Sm}/k_{\mathrm{Nis}}$.

9.2 Sheaves on a Site

To define $\mathcal{A}$-valued sheaves for some value-category $\mathcal{A}$, we need to be able to state the sheaf axiom, so we require that $\mathcal{A}$ admits arbitrary products (indexed by sets). Recall that the *equalizer* of two morphisms $f, f' : A \to B$ in $\mathcal{A}$, is a morphism $i : A_0 \to A$ which is universal for morphisms $g : Z \to A$ such that $fg = f'g$, i.e., there exists a unique morphism $\phi : Z \to A_0$ with $g = i\phi$. If $i : A_0 \to A$ is the equalizer of f and f', we say the sequence

$$A_0 \xrightarrow{\ i\ } A \underset{f'}{\overset{f}{\rightrightarrows}} B \text{ is } exact.$$

Remarks 9.6 *(1) The equalizer of f and f' is the same is the projective limit over the category $I := A \overset{f}{\underset{f'}{\rightrightarrows}} B$ of the evident inclusion functor $I \to A$.*

(2) If A is an abelian category, then $A_0 \overset{i}{\longrightarrow} A \overset{f}{\underset{f'}{\rightrightarrows}} B$ is exact if and only if $0 \to A_0 \overset{i}{\longrightarrow} A \overset{f-f'}{\longrightarrow} B$ is exact in the usual sense.

If A is as above, S an A-valued presheaf on C and $\{f_\alpha : U_\alpha \to X \mid \alpha \in A\} \in \mathrm{Cov}_\tau(X)$ for some $X \in C$, we have the "restriction" morphisms

$$f_\alpha^* : S(X) \to S(U_\alpha)$$
$$p_{1,\alpha,\beta}^* : S(U_\alpha) \to S(U_\alpha \times_X U_\beta)$$
$$p_{2,\alpha,\beta}^* : S(U_\beta) \to S(U_\alpha \times_X U_\beta).$$

Taking products, we have the diagram in A

$$S(X) \xrightarrow{\ \Pi f_\alpha^*\ } \prod_\alpha S(U_\alpha) \overset{\Pi\, p_{1,\alpha,\beta}^*}{\underset{\Pi\, p_{2,\alpha,\beta}^*}{\rightrightarrows}} \prod_{\alpha.\beta} S(U_\alpha \times_X U_\beta). \qquad (10)$$

Definition 9.7 *Let A be a category having arbitrary (small) products, and let τ be a Grothendieck pre-topology on a small category C. An A-valued presheaf S on C is a sheaf for τ if for each covering family $\{f_\alpha : U_\alpha \to X\} \in \mathrm{Cov}_\tau$, the sequence (10) is exact. The category $Shv_\tau^A(C)$ of A-valued sheaves on C for τ is the full subcategory of $PreShv^A(C)$ with objects the sheaves.*

Remark 9.8 *For the examples discussed in 9.4, we use the following notation: If X is a topological space, we write $Shv^A(X)$ for $Sh_X^A(Op(X))$. For X a finite type k-scheme, we write $Sh_\tau^A(X)$ for $Sh_\tau^A(X_\tau)$, where $\tau = \mathrm{Zar}, \text{ét}$, etc. We use a similar notation, $PreShv^A(X)$ (X a topological space) or $PreShv_\tau^A(X)$ (X a finite type k-scheme, $\tau = \mathrm{Zar}, \text{ét}$, etc.) for the respective presheaf categories. In case $A = \mathbf{Sets}$, we sometimes omit the A from the notation.*

For the indiscrete topology on C, we have

$$Shv_{ind}^A(C) = PreShv^A(C)$$

so the presheaf category is also a category of sheaves.

Remark 9.9 *Suppose that our category C has an initial object $\emptyset$ and finite coproducts; we also suppose that the value category A has finite products. A presheaf $P : C^{\mathrm{op}} \to A$ is called* additive *if the canonical map*

$$P\left(\coprod_{i=1}^n X_i\right) \to \prod_{i=1}^n P(X_i)$$

induced by the inclusions $X_i \to \amalg_i X_i$ is an isomorphism; in particular $P(\emptyset)$ is isomorphic to the initial object of $\mathcal{A}$.

Now suppose we have a topology τ on $\mathcal{C}$ such that the collection of inclusions $\{X_i \to \amalg_{i=1}^n X_i \mid i = 1, \ldots, n\}$ is in $\mathrm{Cov}_\tau(\amalg_i X_i)$. Then every sheaf is additive (as a presheaf). Thus, if we wish to understand the τ-sheaves on $\mathcal{C}$, we can restrict our attention to the sheafification of additive presheaves.

An Example: Sheaves on Spec $k_{\text{ét}}$

Let k be a field. The category underlying the étale site Spec $k_{\text{ét}}$ is the opposite of the category of finite, separable k-algebras by sending $k \to A$ to Spec $A \to$ Spec k; each such k-algebra A is a finite product, $A = \prod_{i=1}^n F_i$, with $k \to F_i$ a finite separable field extension, and hence Spec $A \to$ Spec k is just the disjoint union of the finite étale maps Spec $F_i \to$ Spec k.

An additive presheaf P on Spec $k_{\text{ét}}$ (with values in **Ab**) is thus a covariant functor from the category of finite separable k-algebras to **Ab**, which sends products to products. P is thus determined by its value on the subcategory of finite separable field extensions $k \to F$. Note that, if $k \to F$ is finite separable extension each automorphism ϕ of F over k is a morphism in the category of extensions, hence acts on $P(F)$.

In this language, the sheaf axiom is equivalent to the following condition: Let $F \to K$ be a Galois extension over k with group $G := \mathrm{Gal}(K/F)$ and with F and K finite separable field extensions of k. Then the natural map

$$P(F) \to P(K)^G$$

is an isomorphism.

In particular, let G_k be the Galois group of k, i.e. $G_k = \mathrm{Gal}(k_{sep}/k)$ where $k \to k_{sep}$ is a fixed separable closure of k. G_k is a profinite group; let M be a continuous G_k-module, where we give M the discrete topology. For each finite separable subextension $k \to F \to k_{sep}$, let $G_F \subset G_k$ be the subgroup fixing F and let

$$P_M(F) := M^{G_F}$$

be the G_F-fixed submodule of M. Sending M to the functor $F \mapsto P_M(F)$ gives an equivalence of the category of sheaves of abelian groups on Spec $k_{\text{ét}}$ with the category of continuous discrete G_k-modules. The inverse functor is defined by writing k_{sep} as the union of fields F_n over a tower of finite separable extensions of k:

$$k = F_0 \subset F_1 \subset \ldots \subset F_n \subset \ldots \subset k_{sep}$$

and sending P to $\lim_n P(F_n)$.

As we shall see below, one can define sheaf cohomology in the setting of sheaves on a Grothendieck site. The case of cohomology of a sheaf of abelian groups on $X_{\text{ét}}$ thus generalizes the classical theory of Galois cohomology, which is the case $X = \mathrm{Spec}\, k$.

Simplicial Presheaves and Sheaves

For motivic homotopy theory one uses presheaves and sheaves of simplicial sets; we explain how our discussion of presheaves and sheaves of sets extends to this setting.

Let Δ be the category with objects the ordered sets $[n] := \{0, \ldots, n\}$ (with the standard order) and with morphisms the order-preserving maps of sets. The category of simplicial sets $\mathbf{sSets}$ is just the presheaf category $PreShv^{\mathbf{Sets}}(\Delta)$, i.e., the category of functors $F : \Delta^{op} \to \mathbf{Sets}$. Thus, for a small category $\mathcal{C}$, we have the identification of the presheaf category $PreShv^{\mathbf{sSets}}(\mathcal{C})$ with the category of presheaves of sets on $\mathcal{C} \times \Delta$. This in turn is the same as the category of simplicial presheaves on $\mathcal{C}$, i.e., functors $\Delta^{op} \to PreShv^{\mathbf{Sets}}(\mathcal{C})$.

Similarly, if we have a topology τ on $\mathcal{C}$, we have the induced topology τ on $\mathcal{C} \times \Delta$, with $\mathrm{Cov}_\tau(X \times [n]) := i_n(\mathrm{Cov}_\tau(X))$ for each $X \in \mathcal{C}$, where $i_n : \mathcal{C} \to \mathcal{C} \times \Delta$ is the inclusion functor $i_n(X) := X \times [n]$, $i_n(f) = f \times \mathrm{id}_{[n]}$. This defines the category of sheaves on $\mathcal{C} \times \Delta$, which is the same as as the category of functors $\Delta^{op} \to Shv_\tau^{\mathbf{Sets}}(\mathcal{C})$, i.e., simplicial sheaves. Concretely, a simplicial presheaf $n \mapsto F_n$ is a simplicial sheaf if and only if each F_n is a sheaf.

All the basic results on presheaves and sheaves of sets extend without change to presheaves and sheaves of simplicial sets via these identifications.

Projective Limits

The left-exactness of projective limits allows one to construct projective limits of sheaves easily:

Lemma 9.10 *Let $F : I \to PreShv^{\mathcal{A}}(\mathcal{C})$ be a functor such that $F(i)$ is a τ-sheaf for all $i \in I$, $\mathcal{A} = \mathbf{Sets}$ or $\mathcal{A} = \mathbf{Ab}$. Then the presheaf $\varprojlim F$ is a τ-sheaf.*

Proof. Take $\{f_\alpha : U_\alpha \to X\}$ in Cov_τ. For each i, the sequence

$$F(i)(X) \xrightarrow{\Pi f_\alpha^*} \prod_\alpha F(i)(U_\alpha) \overset{\Pi p_{1,\alpha,\beta}^*}{\underset{\Pi p_{2,\alpha,\beta}^*}{\rightrightarrows}} \prod_{\alpha.\beta} F(i)(U_\alpha \times_X U_\beta).$$

is exact. The left-exactness of $\varprojlim$ in $\mathcal{A}$ and the definition of the presheaf limit implies (see remark 9.6) that

$$(\varprojlim F)(X) \xrightarrow{\Pi f_\alpha^*} \prod_\alpha (\varprojlim F)(U_\alpha) \overset{\Pi p_{1,\alpha,\beta}^*}{\underset{\Pi p_{2,\alpha,\beta}^*}{\rightrightarrows}} \prod_{\alpha.\beta} (\varprojlim F)(U_\alpha \times_X U_\beta).$$

is exact, whence the result.

This yields

Proposition 9.11 *Let C be a small category and let τ be a Grothendieck pre-topology on C. Take $\mathcal{A} = \mathbf{Sets}$ or $\mathcal{A} = \mathbf{Ab}$. Then*

1. *small projective limits exist in $Shv_\tau^{\mathcal{A}}(C)$.*
2. *$(F : I \to Shv_\tau^{\mathcal{A}}(C)) \mapsto \varprojlim F$ is left-exact.*

Sheafification

Let $i : Shv_\tau^{\mathcal{A}}(C) \to PreShv^{\mathcal{A}}(C)$ be the inclusion functor. A modification of the sheafification construction for sheaves on a topological space yields:

Theorem 9.12 *For $\mathcal{A} = \mathbf{Sets}, \mathbf{Ab}$, the inclusion i admits a left adjoint a_τ : $PreShv^{\mathcal{A}}(C) \to Shv_\tau^{\mathcal{A}}(C)$.*

The proof proceeds in a number of steps. To begin, a presheaf P is called *separated* (for the pre-topology τ) if for each $\{f_\alpha : U_\alpha \to X\}$ in Cov_τ, the map

$$P(X) \to \prod_\alpha P(U_\alpha)$$

is a monomorphism. We first construct a functor $P \mapsto P^+$ on presheaves with the property that P^+ is separated for each presheaf P.

Definition 9.13 *Let $\mathcal{U} := \{f_\alpha : U_\alpha \to X \mid \alpha \in A\}$ and $\mathcal{V} := \{g_\beta : V_\beta \to X \mid \beta \in B\}$ be in $\mathrm{Cov}_\tau(X)$. We say that $\mathcal{V}$ is a refinement of $\mathcal{U}$ if there exists a map of sets $r : B \to A$ and morphisms $\phi_\beta : V_\beta \to U_{r(\beta)}$ in C for each $\beta \in B$, such that $g_\beta = f_{r(\beta)} \circ \phi_\beta$ for all β. The pair $\rho := (r, \{\phi_\beta, \beta \in B\})$ is called a refinement mapping, written*

$$\rho : \mathcal{V} \to \mathcal{U}.$$

Let $\mathcal{U} = \{f_\alpha : U_\alpha \to X \mid \alpha \in A\}$ be in $\mathrm{Cov}_\tau(X)$. For a presheaf P, define $P_X(\mathcal{U})$ by the exactness of

$$P_X(\mathcal{U}) \longrightarrow \prod_\alpha P(U_\alpha) \underset{\prod p_{2,\alpha,\alpha'}^*}{\overset{\prod p_{1,\alpha,\alpha'}^*}{\rightrightarrows}} \prod_{\alpha.\alpha'} P(U_\alpha \times_X U_{\alpha'}).$$

$P_X(\mathcal{U})$ exists since projective limits exist in **Sets** and in **Ab**. Since $f_\alpha \circ p_{1,\alpha,\alpha'} = f_{\alpha'} \circ p_{2,\alpha,\alpha'}$, the universal property of the equalizer gives us a canonical map

$$\epsilon_\mathcal{U} : P(X) \to P_X(\mathcal{U}).$$

Similarly, each refinement mapping $\rho = (r, \{\phi_\beta\}) : \mathcal{V} \to \mathcal{U}$ gives a commutative diagram

$$P(X) \xrightarrow{\epsilon_{\mathcal{U}}} P_X(\mathcal{U})$$

with $\epsilon_{\mathcal{V}}$ going down-left and ρ^* going down to $P_X(\mathcal{V})$

where ρ^* is induced by the commutative diagram

$$
\begin{array}{ccccc}
P_X(\mathcal{U}) & \longrightarrow & \prod_\alpha P(U_\alpha) & \underset{\prod p^*_{2,\alpha,\alpha'}}{\overset{\prod p^*_{1,\alpha,\alpha'}}{\rightrightarrows}} & \prod_{\alpha.\alpha'} P(U_\alpha \times_X U_{\alpha'}) \\
\Big\downarrow \rho^* & & \Big\downarrow \pi & & \Big\downarrow \pi \\
 & & \prod_\beta P(U_{r(\beta)}) & \underset{\prod p^*_{2,r(\beta),r(\beta')}}{\overset{\prod p^*_{1,r(\beta),r(\beta')}}{\rightrightarrows}} & \prod_{\beta,\beta'} P(U_{r(\beta)} \times_X U_{r(\beta')}) \\
 & & \Big\downarrow \prod_\beta \phi^*_\beta & & \Big\downarrow \prod_{\beta,\beta'} \phi^*_\beta \times \phi^*_{\beta'} \\
P_X(\mathcal{V}) & \longrightarrow & \prod_\beta P(V_\beta) & \underset{\prod p^*_{2,\beta,\beta'}}{\overset{\prod p^*_{1,\beta,\beta'}}{\rightrightarrows}} & \prod_{\beta.\beta'} P(V_\beta \times_X V_{\beta'}),
\end{array}
$$

and where the maps π are the respective projections.

Lemma 9.14 *Suppose we have two refinement mappings* $\rho, \rho' : \mathcal{V} \to \mathcal{U}$. *Then* $\rho^* = \rho'^*$.

Proof. Write $\rho = (r, \{\phi_\beta\})$, $\rho' = (r', \{\phi'_\beta\})$. For each β. let $\psi_\beta : V_\beta \to U_{r(\beta)} \times_X U_{r'(\beta)}$ be the map $(\phi_\beta, \phi'_\beta)$. If $\prod_\alpha x_\alpha$ is in $P_X(\mathcal{U}) \subset \prod_\alpha P(U_\alpha)$ then

$$
\begin{aligned}
\phi^*_\beta(x_{r(\beta)}) &= \psi^*_\beta \circ p^*_{1,r(\beta),r'(\beta)}(x_{r(\beta)}) \\
&= \psi^*_\beta \circ p^*_{2,r(\beta),r'(\beta)}(x_{r'(\beta)}) \\
&= \phi'^*_\beta(x_{r'(\beta)}).
\end{aligned}
$$

Now let $Cov_\tau(X)$ be the category with objects $\mathrm{Cov}_\tau(X)$ and a unique morphism $\mathcal{V} \to \mathcal{U}$ if there exists a refinement mapping $\rho : \mathcal{V} \to \mathcal{U}$. We have defined the functor $P_X : Cov_\tau(X)^{\mathrm{op}} \to \mathcal{A}$ sending $\mathcal{U}$ to $P_X(\mathcal{U})$ and $\mathcal{V} \to \mathcal{U}$ to ρ^* for any choice of a refinement mapping $\rho : \mathcal{V} \to \mathcal{U}$. For later use, we record the structure of the category $Cov_\tau(X)$:

Lemma 9.15 $Cov_\tau(X)^{\mathrm{op}}$ *is a small filtering category.*

Proof. Since $\mathrm{Hom}_{Cov_\tau(X)}(\mathcal{V}, \mathcal{U})$ is either empty or has a single element, properties L1 and L2 follow from L3. If $\mathcal{U} = \{f_\alpha : U_\alpha \to X\}$ and $\mathcal{V} = \{g_\beta : V_\beta \to X\}$ are in $\mathrm{Cov}_\tau(X)$, then $\{U_\alpha \times_X V_\beta \to X\}$ is a common refinement, verifying L3.

We set

$$P^+(X) := \varinjlim_{Cov_\tau(X)^{\mathrm{op}}} P_X.$$

The maps $\epsilon_{\mathcal{U}} : P(X) \to P_X(\mathcal{U})$ define the map $\epsilon_X : P(X) \to P^+(X)$.

Let $g : Y \to X$ be a morphism in $\mathcal{C}$. If $\mathcal{U} = \{f_\alpha : U_\alpha \to X\}$ is in $\mathrm{Cov}_\tau(X)$, we have the covering family $g^*\mathcal{U} := \{p_2 : U_\alpha \times_X Y \to Y\}$ in $\mathrm{Cov}_\tau(Y)$. The operation $\mathcal{U} \mapsto g^*\mathcal{U}$ respects refinement, giving the functor $\hat{g}^* : \mathrm{Cov}_\tau(X) \to \mathrm{Cov}_\tau(Y)$, with $\widehat{(gh)}^*$ canonically isomorphic to $\hat{h}^*\hat{g}^*$. If $\mathcal{U} = \{f_\alpha : U_\alpha \to X\}$ is in $\mathrm{Cov}_\tau(Y)$, pull-back by the projections $U_\alpha \times_X Y \to U_\alpha$ defines the map $g^*_{\mathcal{U}} : P_X(\mathcal{U}) \to P_Y(\hat{g}^*\mathcal{U})$. The maps $g^*_{\mathcal{U}}$ define a natural transformation $g^*_? : P_X \to P_Y \circ \hat{g}^*$. Thus, we have the map on the limits

$$g^* := (\hat{g}^*)_* \circ (g^*_?)_* : P^+(X) \to P^+(Y).$$

With these pull-back maps, we have defined a presheaf P^+ on $\mathcal{C}$; the maps ϵ_X define the map of presheaves $\epsilon_P : P \to P^+$.

If $f : P \to Q$ is a morphism of presheaves, f induces in the evident manner a natural transformation $f_X : P_X \to Q_X$ and hence a map on the limits $f^+(X) : P^+(X) \to Q^+(X)$. The maps $f^+(X)$ define a morphism of presheaves $f^+ : P^+ \to Q^+$, compatible with the maps ϵ_P, ϵ_Q. Thus we have define a functor $^+ : PreShv^{\mathcal{A}}(\mathcal{C}) \to PreShv^{\mathcal{A}}(\mathcal{C})$ and a natural transformation $\epsilon : \mathrm{id} \to {}^+$.

Lemma 9.16 *Let P be a presheaf on $\mathcal{C}$ with values in* **Sets**, **Ab**.

1. *P^+ is a separated presheaf.*
2. *If P is separated, then P^+ is a sheaf*
3. *If P is a sheaf, then $\epsilon_P : P \to P^+$ is an isomorphism.*

Proof. For (1), take $\{f_\alpha : U_\alpha \to X\}$ in $\mathrm{Cov}_\tau(X)$. Since $\mathrm{Cov}_\tau(X)$ is filtering, each element $x \in P^+(X)$ is represented by an $x_\mathcal{V} \in P_X(\mathcal{V})$ for some $\mathcal{V} \in \mathrm{Cov}_\tau(X)$, and $x_\mathcal{V}$, $x_{\mathcal{V}'}$ represent the same element in $P^+(X)$ if and only if there is a common refinement $\rho : \mathcal{W} \to \mathcal{V}$, $\rho' : \mathcal{W} \to \mathcal{V}'$ with $\rho^*(x_\mathcal{V}) = \rho'^*(x_{\mathcal{V}'})$ in $P(\mathcal{W})$. Similar remarks describe $P^+(U_\alpha)$.

Now, if for $x, x' \in P^+(X)$ have $f^*_\alpha(x) = f^*_\alpha(x')$ for all α, choose a covering family $\mathcal{V}$ and elements $x_\mathcal{V}, x'_\mathcal{V} \in P_X(\mathcal{V})$ representing x, x'. Replacing $\mathcal{V}$ and $\mathcal{U}$ with a common refinement, we may assume that $\mathcal{V} = \mathcal{U}$. Write

$$x_\mathcal{U} = \prod_\alpha x_\alpha \in P(U_\alpha),$$

$$x'_\mathcal{U} = \prod_\alpha x'_\alpha \in P(U_\alpha).$$

The element $f^*_\alpha(x_\mathcal{U})$ is represented by the collection $\prod_{\alpha'} p^*_{1\alpha'}(x_{\alpha'})$ in $P_{U_\alpha}(\mathcal{U} \times_X U_\alpha)$, where $\mathcal{U} \times_X U_\alpha$ is the covering family $\{U_{\alpha'} \times_X U_\alpha \to U_\alpha\}$ of U_α. The diagonal $U_\alpha \to U_\alpha \times_X U_\alpha$ gives the refinement $\{\mathrm{id}\} \to \mathcal{U} \times_X U_\alpha$, so $f^*_\alpha(x_\mathcal{U})$ is also represented by $x_\alpha \in P(U_\alpha)$.

The identity $f^*_\alpha(x_\mathcal{U}) = f^*_\alpha(x'_\mathcal{U})$ thus is equivalent to the assertion that there is a covering family $\mathcal{V}_\alpha = \{g_{\alpha\beta} : V_{\alpha\beta} \to U_\alpha\}$ in $\mathrm{Cov}_\tau(U_\alpha)$ such that $g^*_{\alpha\beta}(x_\alpha) =$

$g_{\alpha\beta}^*(x_\alpha')$ for all β. Replacing $\mathcal{U}$ with the covering family $\{f_\alpha \circ g_{\alpha\beta} : V_{\alpha\beta} \to X\}$, we may assume that $\mathcal{V}_\alpha$ is the identity covering family, i.e. that $x_\alpha = x_\alpha'$ in $P(U_\alpha)$. But then $x = x'$, as desired.

For (2), suppose that P is separated, take $\{f_\alpha : U_\alpha \to X\}$ in $\mathrm{Cov}_\tau(X)$, and suppose we have elements $x_\alpha \in P^+(U_\alpha)$ with

$$p_{1,\alpha,\alpha'}^*(x_\alpha) = p_{2,\alpha,\alpha'}^*(x_{\alpha'}) \text{ in } P^+(U_\alpha \times_X U_{\alpha'})$$

for all α, α'. We need to show that there exists an element $x \in P^+(X)$ with $f_\alpha^*(x) = x_\alpha$, as the injectivity of $P^+(X) \to \prod_\alpha P^+(U_\alpha)$ follows from (1).

The element x_α is represented by a collection

$$\prod_\beta x_{\alpha\beta} \in P_{U_\alpha}(\mathcal{V}_\alpha) \subset \prod_\beta P(V_{\alpha\beta})$$

for some $\mathcal{V}_\alpha := \{V_{\alpha\beta} \to U_\alpha\} \in \mathrm{Cov}_\tau(U_\alpha)$. Since P is separated, the relation $p_{1,\alpha,\alpha'}^*(x_\alpha) = p_{2,\alpha,\alpha'}^*(x_{\alpha'})$ in $P^+(U_\alpha \times_X U_{\alpha'})$ implies the relation

$$p_{1,\alpha\beta,\alpha'\beta'}^*(x_{\alpha\beta}) = p_{2,\alpha\beta,\alpha'\beta'}^*(x_{\alpha'\beta'})$$

in $P(V_{\alpha\beta} \times_X V_{\alpha'\beta'})$. This in turn implies that the collection $\prod_{\alpha,\beta} x_{\alpha\beta} \in \prod_{\alpha,\beta} P(V_{\alpha\beta})$ is in the subset $P_X(\mathcal{V})$, where $\mathcal{V}$ is the covering family $\{V_{\alpha\beta} \to X\}$. This yields the desired element of $P^+(X)$.

For (3), let $\mathcal{U}$ be in $\mathrm{Cov}_\tau(X)$ and suppose P is a τ-sheaf. Then $P(X) \to P(\mathcal{U})$ is an isomorphism, hence we have an isomorphism on the limit $\epsilon_P(X) :$ $P(X) \to P^+(X)$.

We define the functor $a_\tau : PreShv^\mathcal{A}(\mathcal{C}) \to Sh_\tau^\mathcal{A}(\mathcal{C})$ by $a_\tau = {}^+ \circ {}^+$. By the lemma just proved, a_τ is well-defined and $a_\tau \circ i$ is naturally isomorphic to the identity on $Sh_\tau^\mathcal{A}(\mathcal{C})$. We call a_τ the *sheafification* functor.

Lemma 9.17 *The sheafification functor* $a_\tau : PreShv^\mathcal{A}(\mathcal{C}) \to Sh_\tau^\mathcal{A}(\mathcal{C})$ *is left-adjoint to the inclusion functor* i.

Proof. Let P be a presheaf, S a sheaf and $f : P \to i(S)$ a morphism of presheaves. Applying a_τ gives the commutative diagram

$$
\begin{array}{ccc}
P & \xrightarrow{\ f\ } & i(S) \\
{\scriptstyle \epsilon_P}\downarrow & & \downarrow{\scriptstyle \epsilon_{i(S)}} \\
a_\tau(P) & \xrightarrow[a_\tau(f)]{} & a_\tau(i(S))
\end{array}
$$

As $\epsilon_{i(S)}$ is an isomorphism, sending f to $\epsilon_{i(S)}^{-1} \circ a_\tau(f)$ gives a natural transformation

$$\theta_{P,S} : \mathrm{Hom}_{PreShv_\tau^\mathcal{A}(\mathcal{C})}(P, i(S)) \to \mathrm{Hom}_{Shv_\tau^\mathcal{A}(\mathcal{C})}(a_\tau(P), S).$$

Similarly sending $g : a_\tau(P) \to S$ to $g \circ \epsilon_P$ gives the inverse to $\theta_{P,S}$.

This completes the proof of theorem 9.12. We record a useful property of sheafification.

Proposition 9.18 *The sheafification functor* $a_\tau : PreShv^{\mathcal{A}}(\mathcal{C}) \to Sh_\tau^{\mathcal{A}}(\mathcal{C})$ *is exact.*

Proof. Since a_τ is a left adjoint, a_τ is right exact. Similarly, the inclusion functor i is left exact. In particular, a projective limit of sheaves is given pointwise:

$$(\varprojlim F : I \to Shv_\tau^{\mathcal{A}}(\mathcal{C}))(X) = \varprojlim F(X) : I \to \mathcal{A}.$$

Since the category $Cov_\tau(X)$ is filtering, the functor $^+$ is exact, so the functor $i \circ a_\tau$ is left exact, hence a_τ is left exact. $\quad\square$

Inductive Limits of Sheaves

We have seen that the theory of projective limits of sheaves is essentially trivial, in that it coincides with the theory of projective limits of presheaves. To construct inductive limits of sheaves, we use the inductive limit of presheaves plus the sheafification functor.

Proposition 9.19 *Let I be a small category, $F : I \to Sh_\tau^{\mathcal{A}}(\mathcal{C})$ a functor. Then $\varinjlim F$ exists and is given by the formula*

$$\varinjlim F = a_\tau(\varinjlim i \circ F).$$

The functor $F \mapsto \varinjlim F$ is right exact; if I is a filtering category, then $F \mapsto \varinjlim F$ is exact.

Proof. We verify the universal property: Take a sheaf S. Then

$$\begin{aligned}
\mathrm{Hom}_{Sh_\tau^{\mathcal{A}}(\mathcal{C})}(a_\tau(\varinjlim i \circ F), S) &= \mathrm{Hom}_{PreSh_\tau^{\mathcal{A}}(\mathcal{C})}(\varinjlim i \circ F, i(S)) \\
&= \varprojlim \mathrm{Hom}_{PreSh_\tau^{\mathcal{A}}(\mathcal{C})}(i \circ F, i(S)) \\
&= \varprojlim \mathrm{Hom}_{Sh_\tau^{\mathcal{A}}(\mathcal{C})}(F, S).
\end{aligned}$$

The exactness assertions follow from the exactness properties of the presheaf inductive limit and the exactness of a_τ. $\quad\square$

Epimorphisms of Sheaves

Since projective limits of sheaves and presheaves agree, a map of sheaves $S' \to S$ is a monomorphism if and only if $f(X) : S'(X) \to S(X)$ is a monomorphism for each X. We now make explicit the condition that a map of sheaves $f : S \to$

S' be an epimorphism. Let 0 be the final object in $\mathcal{A}$ and $*$ the sheafification of the constant presheaf 0 with value 0. It is easy to see that f is an epimorphism if and only if the canonical map $S'/S \to *$ is an isomorphism, where the quotient sheaf S'/S is defined as the inductive limit over the category I

$$
\begin{array}{ccc}
S & \xrightarrow{\ f\ } & S' \\
\downarrow{\scriptstyle \pi} & & \\
* & &
\end{array}
$$

of the evident inclusion functor $I \to Shv_\tau^{\mathcal{A}}(\mathcal{C})$. Noting that the sheaf inductive limit is the sheafification of the presheaf inductive limit we see that S'/S is $*$ if and only if the presheaf $i(S')/i(S)$ has $*$ as sheafification. Since P^+ is separated for every presheaf P, we see that $a_\tau(i(S')/i(S)) = *$ if and only if $(i(S')/i(S))^+ \to 0^+ = *$ is an isomorphism. This in turn yields the criterion:

Proposition 9.20 *A map of sheaves $f : S \to S'$ is an epimorphism if and only if, for each $X \in \mathcal{C}$, and each $x \in S'(X)$, there exists a $\{g_\alpha : U_\alpha \to X\} \in \mathrm{Cov}_\tau(X)$ and $y_\alpha \in S(U_\alpha)$ with $f(y_\alpha) = g_\alpha^*(x)$.*

Functoriality

Let $f : \mathcal{C} \to \mathcal{C}'$ be a functor of small categories, where $\mathcal{C}$ and $\mathcal{C}'$ are endowed with pre-topologies τ, τ', respectively. f is called *continuous* if for each $\{g_\alpha : U_\alpha \to X\} \in \mathrm{Cov}_\tau$, $\{f(g_\alpha) : f(U_\alpha) \to f(X)\}$ is in $\mathrm{Cov}_{\tau'}$. Thus, if f is continuous, the presheaf pull-back $f^p : PreShv^{\mathcal{A}}(\mathcal{C}') \to PreShv^{\mathcal{A}}(\mathcal{C})$ restricts to give the sheaf pull-back $f^s : Shv_{\mathcal{A}}^{\tau'}(\mathcal{C}') \to Shv_\tau^{\mathcal{A}}(\mathcal{C})$, i.e.

$$
i \circ f^s = f^p \circ i'
$$

where i and i' are the respective inclusions.

We define the sheaf push-forward by $f_s := a_{\tau'} \circ f_p \circ i$.

Proposition 9.21 *Let $f : (\mathcal{C}, \tau) \to (\mathcal{C}', \tau')$ be a continuous functor of small categories with pre-topologies. Then*

1. f_s is left-adjoint to f^s
2. f^s is left exact and f_s is right exact.

Proof. The proof of adjointness is formal:

$$
\begin{aligned}
\mathrm{Hom}_{Shv_{\tau'}^{\mathcal{A}}(\mathcal{C}')}(f_s(S), T) &= \mathrm{Hom}_{Shv_{\tau'}^{\mathcal{A}}(\mathcal{C}')}(a_{\tau'} \circ f_p \circ i(S), T) \\
&= \mathrm{Hom}_{PreShv^{\mathcal{A}}(\mathcal{C}')}(f_p \circ i(S), i'(T)) \\
&= \mathrm{Hom}_{PreShv^{\mathcal{A}}(\mathcal{C})}(i(S), f^p \circ i'(T)) \\
&= \mathrm{Hom}_{PreShv^{\mathcal{A}}(\mathcal{C})}(i(S), i \circ f^s(T)) \\
&= \mathrm{Hom}_{Shv_\tau^{\mathcal{A}}(\mathcal{C})}(S, f^s(T)).
\end{aligned}
$$

The exactness assertions follow from the adjoint property.

Example 9.22 *We consider the case of a continuous map $f : X \to Y$ of topological spaces. Let $F : \mathrm{Op}(Y) \to \mathrm{Op}(X)$ be the inverse image functor $F(V) = f^{-1}(V)$, which is clearly continuous with respect to the topologies X, Y on $\mathrm{Op}(X), \mathrm{Op}(Y)$. The functor $F^s : Sh(X) \to Sh(Y)$ is usually denoted f_* and $F_s : Sh(Y) \to Sh(X)$ is denoted f^*.*

More generally, if τ is a pre-topology on $\mathcal{C}$ and X is an object in $\mathcal{C}$, let X_τ be the full subcategory of the category of morphisms to X, $\mathcal{C}/X$, with objects the morphisms $f : U \to X$ which occur as a member of a covering family of X. Clearly τ restricts to a pre-topology on X_τ; we write X_τ for the category with this pre-topology. If $f : X \to Y$ is a morphism in $\mathcal{C}$ then sending $f : U \to Y$ to $p_2 : U \times_Y X \to X$ is a well-defined continuous functor

$$f^{-1} : Y_\tau \to X_\tau.$$

We denote $(f^{-1})^s$ by f_* and f_s^{-1} by f^*.

For example, if X is a k-scheme of finite type, and $\tau = \mathrm{Zar}$, ét, etc., on $\mathbf{Sch}_k$, then X_τ agrees with our definition of X_τ given in example 9.4.

Generators and Abelian Structure

Just as for presheaves, we can use the inclusion functor $i_X : * \to \mathcal{C}$ corresponding to an object $X \in \mathcal{C}$ to construct generators for the sheaf category. Let $\chi_X = i_{X_s}(0) \in Shv_\tau^{\mathbf{Sets}}(\mathcal{C})$ and $\mathbb{Z}_X = i_{X_s}(\mathbb{Z}) \in Shv_\tau^{\mathbf{Ab}}(\mathcal{C})$. Since $\mathrm{Hom}_{Shv_\tau^{\mathbf{Sets}}(\mathcal{C})}(\chi_X, S) = S(X)$, $\mathrm{Hom}_{Shv_\tau^{\mathbf{Ab}}(\mathcal{C})}(\mathbb{Z}_X, S) = S(X)$,, we have

Proposition 9.23 *Let τ be a pre-topology on a small category $\mathcal{C}$. Then $\{\chi_X \mid X \in \mathcal{C}\}$ forms a set of generators for $Shv_\tau^{\mathbf{Sets}}(\mathcal{C})$ and $\{\mathbb{Z}_X \mid X \in \mathcal{C}\}$ forms a set of generators for $Shv_\tau^{\mathbf{Ab}}(\mathcal{C})$.*

The category $Shv_\tau^{\mathbf{Ab}}(\mathcal{C})$ is an abelian category: for a morphism $f : S \to T$, $\ker f$ is just the presheaf kernel (since i is left-exact, this is automatically a sheaf). The sheaf cokernel is given by

$$\mathrm{coker}^s f = a_\tau(\mathrm{coker}^p f).$$

One easily checks that this formula does indeed give the categorical cokernel and that the natural map $\ker(\mathrm{coker} f) \to \mathrm{coker}(\ker f)$ is an isomorphism. Just as for presheaves, the existence of a set of generators, and the exactness of filtered inductive limits gives us

Proposition 9.24 *The abelian category $Shv_\tau^{\mathbf{Ab}}(\mathcal{C})$ has enough injectives.*

Cohomology

Let $\mathcal{A}, \mathcal{B}$ be abelian categories, $f : \mathcal{A} \to \mathcal{B}$ an additive functor. If $\mathcal{A}$ has enough injectives, then each object M in $\mathcal{A}$ admits an injective resolution

$$0 \to M \to I^0 \to \ldots \to I^n \to \ldots .$$

The *right-derived functors* $R^q f$ of f are then defined (see e.g. [13]) and are given by the formula

$$R^q f(M) = H^q(f(I^0) \to f^{(}I^1) \to \ldots \to f(I^n) \to \ldots).$$

The $R^q f$ form a *cohomological functor* in that each short exact sequence $0 \to M' \to M \to M'' \to 0$ in $\mathcal{A}$ yields a long exact sequence

$$0 \to R^0 f(M') \to R^0 f(M) \to R^0 f(M'') \to R^1 f(M') \to \ldots$$
$$\to R^n f(M') \to R^n f(M) \to R^n f(M'') \to R^{n+1} f(M') \to \ldots$$

natural in the exact sequence. If f is left exact, then $R^0 f = f$. For example, the nth right-derived functor of the left exact functor

$$N \mapsto \mathrm{Hom}_{\mathcal{A}}(M, N)$$

is denoted $\mathrm{Ext}^n_{\mathcal{A}}(M, -)$, and $\mathrm{Ext}^0_{\mathcal{A}}(M, N) = \mathrm{Hom}_{\mathcal{A}}(M, N)$.

We apply this to sheaves of abelian groups on $(\mathcal{C}, \tau)$. Take $X \in \mathcal{C}$ and consider the left-exact functor

$$i^s_X : Shv^{\mathbf{Ab}}_\tau(X) \to Shv^{\mathbf{Ab}}(*) = \mathbf{Ab}$$

This is just the functor $S \mapsto S(X)$. We define the *cohomology* $H^n(X, S)$ by

$$H^n(X, S) := R^n i^s_X(S).$$

As $i^s_X(S) = \mathrm{Hom}_{\mathbf{Ab}}(\mathbb{Z}, i^s_X(X)) = \mathrm{Hom}_{Shv^{\mathbf{Ab}}_\tau(\mathcal{C})}(\mathbb{Z}_X, S)$, we have another interpretation of $H^n(X, S)$:

$$H^n(X, S) = \mathrm{Ext}^n_{Shv^{\mathbf{Ab}}_\tau(\mathcal{C})}(\mathbb{Z}_X, S).$$

In the same way, one can take the right-derived functors of the left exact functor $F^s : Shv^{\mathbf{Ab}}_{\tau'}(\mathcal{C}') \to Shv^{\mathbf{Ab}}_\tau(\mathcal{C})$ for any continuous functor $F : (\mathcal{C}, \tau) \to (\mathcal{C}'\tau')$. Taking $F = f^{-1}$ for $f : Y \to X$ a continuous map of topological spaces, or the fiber product functor f^{-1} for $f : Y \to X$ a morphism in $\mathbf{Sch}_k$ (see example 9.22), we have $F^s = f_*$, giving us the right-derived functors $R^n f_*$ of f_*.

Example 9.25 Étale cohomology. *Taking the site $X_{\text{ét}}$, the above discussion gives us the étale cohomology groups $H^q_{\text{ét}}(X, \mathcal{F})$ for $\mathcal{F}$ a sheaf of abelian groups on $X_{\text{ét}}$. Similarly, for a morphism of schemes $f : Y \to X$ and a sheaf $\mathcal{F}$ on $Y_{\text{ét}}$, we have the higher direct image sheaves on $X_{\text{ét}}$, $R^q f_* \mathcal{F}$. As mentioned in section 9.2, in the case $X = \mathrm{Spec}\, k$, k a field, we have the canonical isomorphism*

$$H^q_{\text{ét}}(\mathrm{Spec}\, k, \mathcal{F}) \cong H^q_{ctn}(G_k, \mathcal{F}(k_{sep})),$$

where k_{sep} is a fixed separable closure of k, G_k is the (profinite) Galois group $\mathrm{Gal}(k_{sep}/k)$ and $\mathcal{F}(k_{sep})$ is the limit of the groups $\mathcal{F}(F_i)$, where

$$k = F_0 \subset F_1 \subset \ldots \subset k_{sep}$$

is any tower of finite separable field extensions of k with $k_{sep} = \cup_i F_i$.

Cofinality

It is often convenient to consider sheaves on a full subcategory $\mathcal{C}_0$ of a given small category $\mathcal{C}$; under certain circumstances, the two categories of sheaves are equivalent.

Definition 9.26 *Let* $i : \mathcal{C}_0 \to \mathcal{C}$ *be a full subcategory of a small category* $\mathcal{C}$, τ *a pre-topology on* $\mathcal{C}$. *If*

1. *each* $X \in \mathcal{C}$ *admits a covering family* $\{f_\alpha : U_\alpha \to X\}$ *with the* U_α *in* $\mathcal{C}_0$,
2. *if* Y *and* Z *are in* $\mathcal{C}_0$, X *is in* $\mathcal{C}$ *and* $Y \to X$, $Z \to X$ *are morphisms such that* $Y \times_X Z$ *exists in* $\mathcal{C}$, *then* $Y \times_X Z$ *is in* $\mathcal{C}_0$,

we say that $\mathcal{C}_0$ *is a* cofinal *subcategory of* $\mathcal{C}$ *for the pre-topology* τ.

If $i : \mathcal{C}_0 \to \mathcal{C}$ is cofinal for τ, we define the *induced* pre-topology τ_0 on $\mathcal{C}_0$, where a covering family $\{f_\alpha : U_\alpha \to X\}$, $X \in \mathcal{C}_0$, is an element of $\mathrm{Cov}_\tau(X)$ with the U_α in $\mathcal{C}_0$. Then $i : (\mathcal{C}_0, \tau_0) \to (\mathcal{C}, \tau)$ is continuous.

Proposition 9.27 *Let* $i : (\mathcal{C}_0, \tau_0) \to (\mathcal{C}, \tau)$ *be a cofinal subcategory of* $(\mathcal{C}, \tau)$ *with induced topology* τ_0. *Then*

$$i^s : Shv_\tau^{\mathcal{A}}(\mathcal{C}) \to Shv_{\tau_0}^{\mathcal{A}}(\mathcal{C}_0)$$

is an equivalence of categories (for $\mathcal{A} = \mathbf{Sets}, \mathbf{Ab}$*)*.

Proof. Let S be a sheaf on $\mathcal{C}$. The adjoint property of i_s and i^s yields the canonical morphism $\theta_S : i_s i^s S \to S$, with $i^s \theta_S : i^s i_s i^s S \to i^s S$ an isomorphism. In addition, for $X \in \mathcal{C}$, we can take a covering family $\{f_\alpha : U_\alpha \to X\}$ with the U_α in $\mathcal{C}_0$. Since $U_\alpha \times_X U_\beta$ is also in $\mathcal{C}_0$, the exactness of

$$S(X) \longrightarrow \prod_\alpha S(U_\alpha) \underset{p_2^*}{\overset{p_1^*}{\rightrightarrows}} \prod_{\alpha,\beta} S(U_\alpha \times_X U_\beta)$$

implies that i^s is a faithful embedding and θ_S is an isomorphism. By the adjoint property again, i^s is fully faithful. As $i^s i_s$ is naturally isomorphic to the identity, the proof is complete.

Example 9.28 *The category of affine* k-*schemes of finite type is cofinal in* $\mathbf{Sch}_k$ *for the topologies* Zar, ét *and* Nis *considered in example 9.4.*

Sub-canonical Topologies

Let $\mathcal{C}$ be a small category. The representable functors $\mathrm{Hom}_{\mathcal{C}}(-, X) = \Xi_X \in PreShv^{\mathbf{Sets}}(\mathcal{C})$ give a ready supply of presheaves of sets. For many interesting topologies these presheaves are already sheaves; such topologies are called *sub-canonical*. For example, all the topologies of example 9.4 are sub-canonical.

The *canonical* topology on $\mathcal{C}$ is the finest sub-canonical topology. One can define the canonical topology explicitly as follows:

A set of morphisms in $\mathcal{C}$, $\mathcal{U} := \{f_\alpha : U_\alpha \to X\}$, is an *effective epimorphism* if for each Y in $\mathcal{C}$, the sequence of Hom-sets

$$\mathrm{Hom}_\mathcal{C}(X,Y) \xrightarrow{\prod f_\alpha^*} \prod_\alpha \mathrm{Hom}_\mathcal{C}(U_\alpha, Y) \overset{\prod p_{1,\alpha,\beta}^*}{\underset{\prod p_{2,\alpha,\beta}^*}{\rightrightarrows}} \prod_{\alpha.\beta} \mathrm{Hom}_\mathcal{C}(U_\alpha \times_X U_\beta, Y)$$

is exact, i.e., if for each $Y \in \mathcal{C}$, the representable presheaf Ξ_Y on $\mathcal{C}$ satisfies the sheaf axiom for the set of morphisms $\mathcal{U}$. $\mathcal{U}$ is called a *universal* effective epimorphism if for each morphism $Z \to X$ in $\mathcal{C}$, the fiber products $U_\alpha \times_X Z$ exist and the family $\{U_\alpha \times_X Z \to Z\}$ is an effective epimorphism. Clearly the universal effective epimorphisms define a pre-topology on $\mathcal{C}$ which agrees with the canonical pre-topology.

If τ is a sub-canonical topology, then the representable presheaves Ξ_Y are already sheaves, hence $\chi_Y := a_\tau(\Xi_Y) = \Xi_Y$. We thus have the functor

$$\chi : \mathcal{C} \to Shv_\tau^{\mathbf{Sets}}(\mathcal{C})$$
$$Y \mapsto \mathrm{Hom}_\mathcal{C}(-, Y) = \chi_Y$$

which by the Yoneda lemma is a fully faithful embedding.

In addition, each sheaf S is a quotient of a coproduct of such sheaves. Indeed, let S be a sheaf, and take $X \in \mathcal{C}$ and $x \in S(X)$. Since $S(X) = \mathrm{Hom}_{Shv_\tau^{\mathbf{Sets}}(\mathcal{C})}(\chi_X, S)$, x uniquely corresponds to a morphism $\phi_x : \chi_X \to S$. Taking the coproduct over all pairs (x, X), $x \in S(X)$, we have the morphism

$$L_0(S) := \coprod_{\substack{(x,X) \\ x \in S(X)}} \chi_X \xrightarrow{\phi := \coprod \phi_x} S$$

which is easily seen to be an epimorphism (even of presheaves). If we let $\mathcal{R}$ be the presheaf with

$$\mathcal{R}(Y) = \{(y, y') \in L_0(S)(Y) \mid \phi(Y)(y) = \phi(Y)(y')\}$$

then $\mathcal{R}$ is also a sheaf; composing the projection $p_i : \mathcal{R} \to L_0(S)$ with the epimorphism $L_0(\mathcal{R}) \to \mathcal{R}$ yields the presentation of S as

$$L_1(S) := L_0(\mathcal{R}) \overset{\phi_1}{\underset{\phi_2}{\rightrightarrows}} L_0(S) \xrightarrow{\phi} S$$

Thus we have presented S as an inductive limit of the representable sheaves χ_X, $X \in \mathcal{C}$.

From this point of view, the sheaf category $Shv_\tau^{\mathbf{Sets}}(\mathcal{C})$ can be viewed as an extension of the original category $\mathcal{C}$, in which arbitrary (small) projective and inductive limits exist, containing $\mathcal{C}$ as a full subcategory, and with the closure of $\mathcal{C}$ under inductive limits being the entire category $Shv_\tau^{\mathbf{Sets}}(\mathcal{C})$.

As these properties are valid for *all* sub-canonical topologies on $\mathcal{C}$, one is entitled to ask why one particular sub-canonical topology is chosen over another. The answer here seems to be more art than science, depending strongly on the properties one wishes to control. For the category of motives, the Nisnevich topology has played a central role, but in questions of arithmetic, such as values of L-functions, the étale topology seems to be a more natural choice, and for issues involving purely geometric properties of schemes, the Zariski topology is often more applicable. In fact, some of the most fundamental questions in the theory of motives and its relation to algebraic geometry and arithmetic can be phrased in terms of the behavior of sheaves under a change of topology.

References

1. **Théorie des topos et cohomologie étale des schémas**. Tome 1. Séminaire de Géométrie Algébrique du Bois-Marie 1963–1964 (SGA 4). Dirigé par M. Artin, A. Grothendieck et J. L. Verdier. Avec la collaboration de N. Bourbaki, P. Deligne et B. Saint-Donat. Lecture Notes in Mathematics, Vol. 269. Springer-Verlag, Berlin-New York, 1972.
2. **Théorie des topos et cohomologie étale des schémas**. Tome 2. Séminaire de Géométrie Algébrique du Bois-Marie 1963–1964 (SGA 4). Dirigé par M. Artin, A. Grothendieck et J. L. Verdier. Avec la collaboration de N. Bourbaki, P. Deligne et B. Saint-Donat. Lecture Notes in Mathematics, Vol. 270. Springer-Verlag, Berlin-New York, 1972.
3. Artin, Michael. **Grothendieck Topologies**, Seminar Notes. Harvard Univ. Dept. of Math., Spring 1962.
4. Atiyah, M. F.; Macdonald, I. G. **Introduction to commutative algebra**. Addison-Wesley, 1969.
5. Griffiths, P.; Harris, J. **Principles of algebraic geometry**. Reprint of the 1978 original. Wiley Classics Library. John Wiley & Sons, Inc., New York, 1994.
6. Grothendieck, A. *Sur quelques points d'algèbre homologique*. Tôhoku Math. J. (2) **9** 1957 119–221.
7. Hartshorne, Robin.**Algebraic geometry**. Graduate Texts in Mathematics, No. 52. Springer-Verlag, New York-Heidelberg, 1977.
8. Hironaka, Heisuke. *Resolution of singularities of an algebraic variety over a field of characteristic zero. I, II*. Ann. of Math. (2) **79** (1964), 109–203; ibid. (2) 79 1964 205–326.
9. Kunz, Ernst, **Introduction to commutative algebra and algebraic geometry**, Birkhäuser, Boston, 1985.
10. Matsumura, Hideyuki. **Commutative algebra**. Second edition. Mathematics Lecture Note Series, 56. Benjamin/Cummings Publishing Co., Inc., Reading, Mass., 1980.
11. Milne, J.S., **ëtale cohomology**. Princeton University Press, Princeton, N.J., 1980

12. Mumford, David. **The red book of varieties and schemes**. Second, expanded edition. Lecture Notes in Mathematics, **1358**. Springer-Verlag, Berlin, 1999.

13. Verdier, J.L. *Catégories triangulées, état 0*. Cohomologie étale. Séminaire de Géométrie Algébrique du Bois-Marie SGA $4\frac{1}{2}$. P. Deligne, avec la collaboration de J. F. Boutot, A. Grothendieck, L. Illusie et J. L. Verdier. pp. 262–311. Lecture Notes in Mathematics, **569**. Springer-Verlag, Berlin-New York, 1977.

14. Zariski, Oscar; Samuel, Pierre. **Commutative algebra**. Vols. 1,2. With the cooperation of I. S. Cohen. Corrected reprinting of the 1958 edition. Graduate Texts in Mathematics, **28**, **29**. Springer-Verlag, New York-Heidelberg-Berlin, 1975.

15. Voevodsky, Vladimir; Suslin, Andrei; Friedlander, Eric. **Cycles, Transfers and Motivic Homology Theories**, Annals of Math. Studies **143**, Princeton Univ. Press. 2000.

Index

144   M. Levine

Voevodsky's Nordfjordeid Lectures: Motivic Homotopy Theory

Vladimir Voevodsky[1], Oliver Röndigs[2], and Paul Arne Østvær[3]

[1] School of Mathematics, Institute for Advanced Study, Princeton, USA
vladimir@math.ias.edu
[2] Fakultät für Mathematik, Universität Bielefeld, Bielefeld, Germany
oroendig@math.uni-bielefeld.de
[3] Department of Mathematics, University of Oslo, Oslo, Norway
paularne@math.uio.no

1 Introduction

Motivic homotopy theory is a new and in vogue blend of algebra and topology. Its primary object is to study algebraic varieties from a homotopy theoretic viewpoint. Many of the basic ideas and techniques in this subject originate in algebraic topology.

This text is a report from Voevodsky's summer school lectures on motivic homotopy in Nordfjordeid. Its first part consists of a leisurely introduction to motivic stable homotopy theory, cohomology theories for algebraic varieties, and some examples of current research problems. As background material, we recommend the lectures of Dundas [Dun] and Levine [Lev] in this volume. An introductory reference to motivic homotopy theory is Voevodsky's ICM address [Voe98]. The appendix includes more in depth background material required in the main body of the text. Our discussion of model structures for motivic spectra follows Jardine's paper [Jar00].

In the first part, we introduce the motivic stable homotopy category. The examples of motivic cohomology, algebraic K-theory, and algebraic cobordism illustrate the general theory of motivic spectra. In March 2000, Voevodsky [Voe02b] posted a list of open problems concerning motivic homotopy theory. There has been so much work done in the interim that our update of the status of these conjectures may be useful to practitioners of motivic homotopy theory.

The second and third author would like to thank Vladimir Voevodsky for helpful discussions concerning the content of this text, and his kind permission to include a sketch proof of Theorem 4.3. The actual wording here, and the responsibility for any misinterpretations, are our own.

2 Motivic Stable Homotopy Theory

In this section, we introduce the motivic stable homotopy category. Although the construction of this category can be carried out for more general base schemes, we shall only consider Zariski spectra of fields.

A final word about precursors: In what follows, we use techniques which are basic in the study of both model categories and triangulated categories. Introductory textbooks on these subjects include [Hov99] and [Nee01].

2.1 Spaces

Let k be a field and consider the category Sm/k of smooth separated schemes of finite type over $\mathrm{Spec}(k)$. From a homotopical point of view, the category Sm/k is intractable because it is not closed under colimits.

The spaces we consider are the objects in the category

$$\mathrm{Spc}(k) \colon = \Delta^{\mathrm{op}} \mathrm{Shv}_{\mathrm{Nis}}(\mathrm{Sm}/k)$$

of Nisnevich sheaves on Sm/k [Lev] with values in simplicial sets [Dun].

We mention two typical types of examples of such sheaves. First, any scheme in Sm/k determines a representable space via the Yoneda embedding. This holds since the Nisnevich topology is sub-canonical [Lev]. Second, any simplicial set can be viewed as a constant Nisnevich sheaf on Sm/k, and also as a constant sheaf in any other Grothendieck topology.

A pointed space consists of a space X together with a map

$$x\colon \mathrm{Spec}(k) \longrightarrow X \, .$$

Here, we consider $\mathrm{Spec}(k)$ as a representable sheaf with constant simplicial structure. Let $\mathrm{Spc}_\bullet(k)$ denote the category of pointed spaces. If X is a space, let X_+ denote the canonically pointed space $X \amalg \mathrm{Spec}(k)$. By adding disjoint base-points, it follows that the forgetful functor from $\mathrm{Spc}_\bullet(k)$ to $\mathrm{Spc}(k)$ has a left adjoint.

It is important to note that the category of pointed spaces has a symmetric monoidal structure: Suppose X and Y are pointed spaces. Then their smash product $X \wedge Y$ is the space associated to the presheaf

$$U \longmapsto X(U) \wedge Y(U) \, .$$

The sheaf represented by the Zariski spectrum $\mathrm{Spec}(k)$ is the terminal presheaf with value the one-point set

$$U \longmapsto * \, .$$

Clearly, this shows that $\mathrm{Spec}(k)_+$ is a unit for the smash product.

Recall that in classical stable homotopy theory, when constructing spectra of pointed simplicial sets one inverts only one suspension coordinate, namely the simplicial circle. This part works slightly differently in the motivic context. An exotic aspect, which turns out to play a pivotal role in the motivic stable homotopy theory, is the use of two radically different suspension coordinates. In order to define the motivic stable homotopy category, we shall make use of bispectra of pointed spaces.

The first of the motivic circles is well-known to topologists: Let $\Delta[n]$ denote the standard simplicial n-simplex [Dun]. Recall that the simplicial circle S^1 is the coequalizer of the diagram

$$\Delta[0] \rightrightarrows \Delta[1] \, .$$

We denote by S_s^1 the corresponding pointed space.

The second motivic circle is well-known to algebraic geometers: Denote by $\mathbb{A}^1 \in \mathrm{Sm}/k$ the affine line. Then the Tate circle S_t^1 is the space $\mathbb{A}^1 \smallsetminus 0$, pointed by the global section given by the identity; this is the underlying scheme of the multiplicative group.

Since pointed spaces $\mathrm{Spc}_\bullet(k)$ acquires a smash product, we may form the n-fold smash products S_s^n and S_t^n of the simplicial circle and the Tate circle. A mixed sphere refers to a smash product of S_s^m and S_t^n.

2.2 The Motivic s-Stable Homotopy Category $\mathbf{SH}_s^{\mathbb{A}^1}(k)$

To invert S_s^1 we shall consider spectra of pointed spaces. This is analogous to the situation with ordinary spectra and the simplicial circle.

Definition 2.1 *An s-spectrum E is a sequence of pointed spaces $\{E_n\}_{n \geq 0}$ together with structure maps*

$$S_s^1 \wedge E_n \longrightarrow E_{n+1} .$$

A map of s-spectra

$$E \longrightarrow E'$$

consists of degree-wise maps of pointed spaces

$$E_n \longrightarrow E'_n$$

which are compatible with the structure maps.

Let $\mathrm{Spt}_s(k)$ denote the category of s-spectra.

Pointed spaces give examples of s-spectra:

Example 2.2 *The s-suspension spectrum of a pointed space X is the s-spectrum $\Sigma_s^\infty X$ with n-th term $S_s^n \wedge X$ and identity structure maps.*

The next step is to define weak equivalences of s-spectra. If $n \geq 1$ and (X, x) is a pointed space, let $\pi_n(X, x)$ denote the sheaf of homotopy groups associated to the presheaf

$$U \longmapsto \pi_n(X(U), x|U) ,$$

where $x|U$ is the image of x in $X(U)$. If $n \geq 2$, this is a Nisnevich sheaf of abelian groups.

The suspension homomorphism for ordinary pointed simplicial sets yields suspension homomorphisms of sheaves of homotopy groups

$$\pi_n(X) \longrightarrow \pi_{n+1}(S_s^1 \wedge X) .$$

If E is an s-spectrum and $m > n$ are integers, consider the sequence

$$\pi_{n+m}(E_m) \longrightarrow \pi_{n+m+1}(S_s^1 \wedge E_m) \longrightarrow \pi_{n+m+1}(E_{m+1}) \longrightarrow \cdots .$$

The sheaves of stable homotopy groups of E are the sheaves of abelian groups

$$\pi_n(E) := \operatorname*{colim}_{m > n} \pi_{n+m}(E_m) .$$

A map between s-spectra E and E' is called an s-stable weak equivalence if for every integer $n \in \mathbb{Z}$, there is an induced isomorphism of sheaves

$$\pi_n(E) \longrightarrow \pi_n(E') .$$

Definition 2.3 *Let* $\mathrm{SH}_s(k)$ *be the category obtained from* $\mathrm{Spt}_s(k)$ *by inverting the s-stable weak equivalences.*

Remark 2.4 *One can show that* $\mathrm{Spt}_s(k)$ *has the structure of a proper simplicial stable model category. The associated homotopy category is* $\mathrm{SH}_s(k)$. *In this model structure, the weak equivalences are the s-stable weak equivalences.*

There is an obvious way to smash an s-spectrum with any pointed space. If we smash with the simplicial circle, the induced simplicial suspension functor Σ_s^1 becomes an equivalence of $\mathrm{SH}_s(k)$: Indeed, an s-spectrum is a Nisnevich sheaf on Sm/k with values in the category Spt of ordinary spectra. Since the simplicial suspension functor is an equivalence of the stable homotopy category SH, it is also an equivalence of $\mathrm{SH}_s(k)$. The term 's-stable' refers to this observation.

Remark 2.5 *There is a canonical functor*

$$\mathrm{Spt} \longrightarrow \mathrm{Spt}_s(k)$$

obtained by considering pointed simplicial sets as pointed spaces. A stable weak equivalence of spectra induces an s-stable weak equivalence of s-spectra, and there is an induced functor

$$\mathrm{SH} \longrightarrow \mathrm{SH}_s(k)$$

between the corresponding homotopy categories.

The basic organizing principle in motivic homotopy theory is to make the affine line contractible. One way to obtain this is as follows: An s-spectrum F is $\mathbb{A}^1$-local if for all $U \in \mathrm{Sm}/k$ and $n \in \mathbb{Z}$, there is a bijection

$$\mathrm{Hom}_{SH_s(k)}(\Sigma_s^\infty U_+, \Sigma_s^n F) \longrightarrow \mathrm{Hom}_{SH_s(k)}(\Sigma_s^\infty (U \times \mathbb{A}^1)_+, \Sigma_s^n F)$$

defined by the projection

$$U \times \mathbb{A}^1 \longrightarrow U .$$

We say that a map

$$E \longrightarrow E'$$

of s-spectra is an $\mathbb{A}^1$-stable weak equivalence if for any $\mathbb{A}^1$-local s-spectrum F, there is a canonically induced bijection

$$\mathrm{Hom}_{\mathrm{SH}_s(k)}(E', F) \longrightarrow \mathrm{Hom}_{\mathrm{SH}_s(k)}(E, F) .$$

Definition 2.6 *Let the motivic s-stable homotopy category* $\mathrm{SH}_s^{\mathbb{A}^1}(k)$ *be the category obtained from* $\mathrm{Spt}_s(k)$ *by inverting the* $\mathbb{A}^1$*-stable weak equivalences.*

The following Lemma is now evident.

Lemma 2.7 *Assume $X \in \mathrm{Sm}/k$. Then the canonical map*

$$X \times \mathbb{A}^1 \longrightarrow X$$

induces an $\mathbb{A}^1$-stable weak equivalence of s-spectra.
In particular, the s-suspension spectrum $\Sigma_s^\infty (\mathbb{A}^1, 0)$ is contractible.

We note that the simplicial suspension functor is an equivalence of $\mathrm{SH}_s^{\mathbb{A}^1}(k)$. The last step in the construction of the motivic stable homotopy category is to invert the Tate circle so that smashing with S_t^1 becomes an equivalence. But first we discuss an unsatisfactory facet of the motivic s-stable homotopy category; part of this is motivation for work in the next section.

In topology, a finite unramified covering map of topological spaces

$$Y \longrightarrow Z \,,$$

induces a transfer map

$$\Sigma^\infty Z_+ \longrightarrow \Sigma^\infty Y_+$$

in the ordinary stable homotopy category. The first algebraic analogue of a covering map is a finite Galois extension of fields, say k'/k. However, in the s-stable homotopy category $\mathrm{SH}_s^{\mathbb{A}^1}(k)$ there is no non-trivial transfer map

$$\Sigma_s^\infty \mathrm{Spec}(k)_+ \longrightarrow \Sigma_s^\infty \mathrm{Spec}(k')_+ \,.$$

This follows by explicit calculations: If E is an s-spectrum and π^s denotes ordinary stable homotopy groups, then

$$\mathrm{Hom}_{\mathrm{SH}_s(k)}(\Sigma_s^\infty S_s^n, E) = \mathrm{Hom}_{\mathrm{SH}}(\Sigma^\infty S^n, E(\mathrm{Spec}(k)))$$
$$= \pi_n^s(E(\mathrm{Spec}(k))) \,.$$

Now, let $X \in \mathrm{Sm}/k$ and suppose that

$$\mathrm{Hom}_{\mathrm{Sm}/k}(\mathbb{A}^1, X) = \mathrm{Hom}_{\mathrm{Sm}/k}(Spec(k), X) \,.$$

This is satisfied if $X = \mathrm{Spec}(k')$, where k' is a field extension of k. Then the above implies an isomorphism

$$\mathrm{Hom}_{\mathrm{SH}_s^{\mathbb{A}^1}(k)}(\Sigma_s^\infty S_s^n, \Sigma_s^\infty X_+) = \mathrm{Hom}_{\mathrm{SH}}(\Sigma^\infty S^n, \Sigma^\infty X(\mathrm{Spec}(k))_+) \,.$$

When $X = \mathrm{Spec}(k')$, note that

$$\mathrm{Hom}_{Sm/k}(Spec(k), Spec(k')) = \emptyset \,.$$

By letting $n = 0$, we find

$$\mathrm{Hom}_{\mathrm{SH}_s^{\mathbb{A}^1}(k)}(\Sigma_s^\infty \mathrm{Spec}(k)_+, \Sigma_s^\infty \mathrm{Spec}(k')_+)$$
$$= \mathrm{Hom}_{\mathrm{SH}}(\Sigma^\infty S^0, \Sigma^\infty \mathrm{Spec}(k')(\mathrm{Spec}(k))_+)$$
$$= \mathrm{Hom}_{\mathrm{SH}}(\Sigma^\infty S^0, \Sigma^\infty *)$$
$$= 0 .$$

Remark 2.8 *It turns out that the existence of transfer maps for finite Galois extensions and the Tate circle being invertible in the homotopy category are closely related issues. Transfer maps are incorporated in Voevodsky's derived category* $\mathrm{DM}^{\mathrm{eff}}_-(k)$ *of effective motivic complexes over k [Voe00b]. There is a canonical Hurewicz map relating the motivic stable homotopy category with* $\mathrm{DM}^{\mathrm{eff}}_-(k)$. *If k is a perfect field, the cancellation theorem [Voe02a] shows that tensoring with the Tate object* $\mathbb{Z}(1)$ *in* $\mathrm{DM}^{\mathrm{eff}}_-(k)$ *induces an isomorphism*

$$\mathrm{Hom}_{\mathrm{DM}^{\mathrm{eff}}_-(k)}(C, D) \xrightarrow{\ \cong\ } \mathrm{Hom}_{\mathrm{DM}^{\mathrm{eff}}_-(k)}\big(C \otimes \mathbb{Z}(1), D \otimes \mathbb{Z}(1)\big) .$$

2.3 The Motivic Stable Homotopy Category SH(k)

The definition of $\mathrm{SH}(k)$ combines the category of s-spectra and the Tate circle. We use bispectra in order to make this precise.

Let $m, n \geq 0$ be integers. An (s, t)-bispectrum E consists of pointed spaces $E_{m,n}$ together with structure maps

$$\sigma_s \colon S_s^1 \wedge E_{m,n} \longrightarrow E_{m+1,n} ,$$

$$\sigma_t \colon S_t^1 \wedge E_{m,n} \longrightarrow E_{m,n+1} .$$

In addition, the structure maps are required to be compatible in the sense that the following diagram commutes.

$$
\begin{array}{ccc}
S_s^1 \wedge S_t^1 \wedge E_{m,n} & \xrightarrow{\ \tau \wedge E_{m,n}\ } & S_t^1 \wedge S_s^1 \wedge E_{m,n} \\
{\scriptstyle S_s^1 \wedge \sigma_t}\Big\downarrow & & \Big\downarrow{\scriptstyle S_t^1 \wedge \sigma_s} \\
S_s^1 \wedge E_{m,n+1} \xrightarrow{\ \sigma_s\ } E_{m+1,n+1} & \xleftarrow{\ \sigma_t\ } & S_t^1 \wedge E_{m+1,n}
\end{array}
$$

Here, τ flips the copies of S_s^1 and S_t^1. There is an obvious notion of maps between such bispectra. An (s, t)-bispectrum can and will be interpreted as a t-spectrum object in the category of s-spectra, that is, a collection of s-spectra

$$E_n := E_{*,n}$$

together with maps of s-spectra induced by the structure maps

$$S_t^1 \wedge E_n \longrightarrow E_{n+1} .$$

We write $\mathrm{Spt}_{s,t}(k)$ for the category of (s,t)-bispectra. If X is a pointed space, let $\Sigma_{s,t}^{\infty} X$ denote the corresponding suspension (s,t)-bispectrum.

If E is an (s,t)-bispectrum and p,q are integers, denote by $\pi_{p,q}(E)$ the sheaf of bigraded stable homotopy groups associated to the presheaf

$$ U \longmapsto \underset{m}{\mathrm{colim}}\ \mathrm{Hom}_{\mathrm{SH}_s^{\mathbb{A}^1}(k)}(S_s^{p-q} \wedge S_t^{q+m} \wedge \Sigma_s^{\infty} U_+, E_m)\ . $$

This expression makes sense for $p < q$, since smashing with S_s^1 yields an equivalence of categories

$$ \mathrm{SH}_s^{\mathbb{A}^1}(k) \longrightarrow \mathrm{SH}_s^{\mathbb{A}^1}(k), \quad E \longmapsto S_s^1 \wedge E\ . $$

Moreover, in the above we assume $q + m \geq 0$.

Definition 2.9 *A map $E \to E'$ of (s,t)-bispectra is a stable weak equivalence if for all $p,q \in \mathbb{Z}$, there is an induced isomorphism of sheaves of bigraded stable homotopy groups*

$$ \pi_{p,q}(E) \longrightarrow \pi_{p,q}(E')\ . $$

We are ready to define our main object of study:

Definition 2.10 *The motivic stable homotopy category* $\mathrm{SH}(k)$ *of k is obtained from* $\mathrm{Spt}_{s,t}(k)$ *by inverting the stable weak equivalences.*

Remark 2.11 *There is an underlying model category structure on* $\mathrm{Spt}_{s,t}(k)$; *the weak equivalences are the stable weak equivalences defined in 2.9. Moreover, this model structure is stable, proper, and simplicial.*

By construction, the suspension functors Σ_s^1 and Σ_t^1 induce equivalences of the motivic stable homotopy category. Hence, analogous to 2.7, we have:

Lemma 2.12 *Assume $X \in \mathrm{Sm}/k$. Then the canonical map*

$$ X \times \mathbb{A}^1 \longrightarrow X $$

induces a stable weak equivalence of suspension (s,t)-bispectra.

In 2.13, we note that maps between suspension spectra in $\mathrm{SH}(k)$ can be expressed in terms of maps in $\mathrm{SH}_s^{\mathbb{A}^1}(k)$. As the proof in the Nisnevich topology hinges on finite cohomological dimension, it is not clear whether there is a similarly general result in the etale topology.

Proposition 2.13 *If $X \in \mathrm{Sm}/k$ and $E \in \mathrm{Spt}_{s,t}(k)$, there is an isomorphism*

$$ \mathrm{Hom}_{\mathrm{SH}(k)}(\Sigma_{s,t}^{\infty} X_+, E) = \underset{n}{\mathrm{colim}}\ \mathrm{Hom}_{\mathrm{SH}_s^{\mathbb{A}^1}(k)}(S_t^n \wedge \Sigma_s^{\infty} X_+, E_n)\ . $$

Proof. If $E = (E_{m,n})$ is some (s,t)-bispectrum, recall that E_n is the s-spectrum defined by the sequence $(E_{0,n}, E_{1,n}, \ldots)$. By model category theory, there exists a fibrant replacement E^f of E in $\mathrm{Spt}_{s,t}(k)$ and isomorphisms

$$\mathrm{Hom}_{\mathrm{SH}(k)}(\Sigma_{s,t}^\infty X_+, E) = \mathrm{Hom}_{\mathrm{Spt}_{s,t}(k)}(\Sigma_{s,t}^\infty X_+, E^f)/_\simeq$$
$$= \mathrm{Hom}_{\mathrm{Spt}_s(k)}(\Sigma_s^\infty X_+, E_0^f)/_\simeq .$$

The relation $\simeq$ is the homotopy relation on maps; in our setting, homotopies are parametrized by the affine line $\mathbb{A}^1$. By using properties of the Nisnevich topology – every scheme has finite cohomological dimension, coverings are generated by so-called upper distinguished squares which will be introduced in Definition 2.18 – one can choose a fibrant replacement E^f so that

$$E_0^f = \operatorname*{colim}_n \left((E_0)^f \longrightarrow \Omega_t((E_1)^f) \longrightarrow \Omega_t(\Omega_t((E_2)^f)) \longrightarrow \cdots \right) .$$

Here, $(E_n)^f$ is a fibrant replacement of E_n in $\mathrm{Spt}_s(k)$, and Ω_t is the right adjoint of the functor

$$S_t^1 \wedge - : \mathrm{Spt}_s(k) \longrightarrow \mathrm{Spt}_s(k) .$$

Since the s-suspension spectra $\Sigma_s^\infty X_+$ and $\Sigma_s^\infty X_+ \wedge \mathbb{A}_+^1$ are both finitely presentable objects in $\mathrm{Spt}_s(k)$, we have

$$\mathrm{Hom}_{\mathrm{Spt}_s(k)}(\Sigma_s^\infty X_+, E_0^f)/_\simeq = \operatorname*{colim}_n \mathrm{Hom}_{\mathrm{Spt}_s(k)}(\Sigma_s^\infty X_+, \Omega_t^n((E_n)^f))/_\simeq .$$

The latter and the isomorphism

$$\mathrm{Hom}_{\mathrm{SH}_s(k)}(\Sigma_s^\infty X_+, \Omega_t^n(E_n)) = \mathrm{Hom}_{\mathrm{SH}_s(k)}(S_t^n \wedge \Sigma_s^\infty X_+, E_n) ,$$

obtained from the adjunction, imply the claimed group isomorphism.   □

Suppose $X, Y \in \mathrm{Sm}/k$. As a special case of 2.13, we obtain an isomorphism between $\mathrm{Hom}_{\mathrm{SH}(k)}(\Sigma_{s,t}^\infty X_+, \Sigma_{s,t}^\infty Y_+)$ and

$$\operatorname*{colim}_m \mathrm{Hom}_{\mathrm{SH}_s^{\mathbb{A}^1}(k)}(S_t^m \wedge \Sigma_s^\infty X_+, S_t^m \wedge \Sigma_s^\infty Y_+) .$$

Remark 2.14 *We will not discuss details concerning the notoriously difficult notion of an adequate smash product for (bi)spectra. Rather than using spectra, one solution is to consider Jardine's category of motivic symmetric spectra [Jar00]. An alternate solution using motivic functors is discussed by Dundas in this volume [Dun].*

For our purposes, it suffices to know that a handicrafted smash product of bispectra induces a symmetric monoidal structure on $\mathrm{SH}(k)$. The proof of this fact is tedious, but straight-forward. The unit for the monoidal structure is the 'sphere spectrum' or the suspension (s,t)-bispectrum $\Sigma_{s,t}^\infty \mathrm{Spec}(k)_+$ of the base scheme k. Moreover, for all $X, Y \in \mathrm{Sm}/k$, there is an isomorphism between the smash product $\Sigma_{s,t}^\infty X_+ \wedge \Sigma_{s,t}^\infty Y_+$ and $\Sigma_{s,t}^\infty (X \times Y)_+$.

Next we summarize the constructions

$$\mathrm{Spc}_\bullet(k) \xrightarrow{\;\Sigma_s^\infty\;} \mathrm{SH}_s^{\mathbb{A}^1}(k) \xrightarrow{\;\Sigma_t^\infty\;} \mathrm{SH}(k) \ .$$

The spaces in the motivic setting are the pointed simplicial sheaves on the Nisnevich site of Sm/k. By using the circles S_s^1 and S_t^1, one defines spectra of pointed spaces as for spectra of simplicial pointed sets. The notion of $\mathbb{A}^1$-stable weak equivalences of s-spectra forces the s-suspension spectrum of $(\mathbb{A}^1, 0)$, the affine line pointed by zero, to be contractible. We use the same class of maps to define the motivic s-stable homotopy category $\mathrm{SH}_s^{\mathbb{A}^1}(k)$.

Our main object of interest, the motivic stable homotopy category $\mathrm{SH}(k)$ is obtained by considering (s, t)-bispectra and formally inverting the class of stable weak equivalences; such maps are defined in terms of Nisnevich sheaves of bigraded homotopy groups.

If

$$\Sigma_{s,t}^\infty \colon \mathrm{Sm}/k \longrightarrow \mathrm{SH}(k)$$

denotes the suspension (s, t)-bispectrum functor, there is a natural equivalence of functors

$$\Sigma_{s,t}^\infty = \Sigma_t^\infty \circ \Sigma_s^\infty \ .$$

Remark 2.15 *Work in progress by Voevodsky suggests yet another construction of* $\mathrm{SH}(k)$. *This uses a theory of framed correspondences; a distant algebraic relative of framed cobordisms, which may have computational advantages.*

Next, we shall specify a triangulated structure on $\mathrm{SH}(k)$. As for a wide range of other examples, this additional structure provides a convenient tool to construct long exact sequences.[4]

The category $\mathrm{Spt}_{s,t}(k)$ is obviously complete and cocomplete: Limits and colimits are formed degree-wise in $\mathrm{Spc}_\bullet(k)$. In particular, there is an induced coproduct $\vee$ in $\mathrm{SH}(k)$. We also claim that the latter is an additive category: Since the simplicial circle S_s^1 is a cogroup object in the ordinary unstable homotopy category, we conclude that every spectrum in $\mathrm{SH}(k)$ is a two-fold simplicial suspension. Thus all objects in $\mathrm{SH}(k)$ are abelian cogroup objects, and the set of maps out of any object is an abelian group.

To define a triangulated category structure on $\mathrm{SH}(k)$, we need to specify a class of distinguished triangles and also a shift functor $[-]$. Suppose that E is an (s, t)-bispectrum. Its shift $E[1]$ is the s-suspension of E. Any map

$$f \colon E \longrightarrow E'$$

of (s, t)-bispectra has an associated cofibration sequence

---

[4] The homotopy categories $\mathrm{SH}_s(k)$ and $\mathrm{SH}_s^{\mathbb{A}^1}(k)$ are also triangulated.

$$E \longrightarrow E' \longrightarrow \mathrm{Cone}(f) \longrightarrow \Sigma_s^1 E \ .$$

The cone of f is defined in terms of a push-out square in $\mathrm{Spt}_{s,t}(k)$, where $\Delta[1]$ is pointed by zero:

$$
\begin{array}{ccc}
E & \xrightarrow{\;E \wedge 0\;} & E \wedge \Delta[1] \\
{\scriptstyle f}\big\downarrow & & \big\downarrow \\
E' & \longrightarrow & \mathrm{Cone}(f)
\end{array}
$$

The map

$$\mathrm{Cone}(f) \longrightarrow \Sigma_s^1 E$$

collapses E' to a point.

A distinguished triangle in $\mathrm{SH}(k)$ is a sequence that is isomorphic to the image of a cofibration sequence in $\mathrm{Spt}_{s,t}(k)$. It follows that any distinguished triangle in $\mathrm{SH}(k)$

$$X \longrightarrow Y \longrightarrow Z \longrightarrow X[1] \ ,$$

induces long exact sequences of abelian groups

$$\cdots [E, X[n]] \longrightarrow [E, Y[n]] \longrightarrow [E, Z[n]] \longrightarrow [E, X[n+1]] \cdots \ ,$$

$$\cdots [Z[n], E] \longrightarrow [Y[n], E] \longrightarrow [X[n], E] \longrightarrow [Z[n-1], E] \cdots \ .$$

Here, $[-, -]$ denotes $\mathrm{Hom}_{\mathrm{SH}(k)}(-, -)$.

The next lemma points out an important class of distinguished triangles.

Lemma 2.16 *If*

$$X \rightarrowtail Y$$

is a monomorphism of pointed spaces, then

$$\Sigma_{s,t}^\infty X \longrightarrow \Sigma_{s,t}^\infty Y \longrightarrow \Sigma_{s,t}^\infty Y/X \longrightarrow \Sigma_{s,t}^\infty X[1]$$

is a distinguished triangle in $\mathrm{SH}(k)$.

Proof. The model structure in Remark 2.11 shows that the canonically induced map

$$\mathrm{Cone}(\Sigma_{s,t}^\infty(X \rightarrowtail Y)) \longrightarrow \Sigma_{s,t}^\infty Y/X$$

is a stable weak equivalence. $\qquad \Box$

Remark 2.17 *Note that 2.16 applies to open and closed embeddings in* Sm/k.

Definition 2.18 *An upper distinguished square is a pullback square in* Sm/k

$$
\begin{array}{ccc}
W & \longrightarrow & V \\
\downarrow & & \downarrow p \\
U & \overset{i}{\longrightarrow} & X
\end{array}
$$

where i *is an open embedding,* p *is an etale map, and*

$$
p|_{p^{-1}(X \smallsetminus U)} \colon p^{-1}(X \smallsetminus U) \longrightarrow (X \smallsetminus U)
$$

induces an isomorphism of reduced schemes.

Any Zariski open covering gives an example of an upper distinguished square: If $X = U \cup V$, let $W = U \cap V$.

The Nisnevich topology is discussed in detail in [Lev]. The generating coverings are of the form

$$
\{ i \colon U \longrightarrow X, p \colon V \longrightarrow X \} .
$$

where i, p and $W = p^{-1}(U)$ form an upper distinguished square.

Since spaces are sheaves in the Nisnevich topology, we get:

Lemma 2.19 *A square of representable spaces which is obtained from an upper distinguished square is a pushout square.*

In the following, we show that upper distinguished squares give examples of distinguished triangles in $\mathrm{SH}(k)$. Part (a) of the next result can be thought of as a generalized Mayer-Vietoris property:

Corollary 2.20 *For an upper distinguished square, the following holds.*

(a) There is a distinguished triangle in $\mathrm{SH}(k)$:

$$
\Sigma_{s,t}^{\infty} W_+ \longrightarrow \Sigma_{s,t}^{\infty} U_+ \vee \Sigma_{s,t}^{\infty} V_+ \longrightarrow \Sigma_{s,t}^{\infty} X_+ \longrightarrow \Sigma_{s,t}^{\infty} W_+[1] .
$$

(b) There is a naturally induced stable weak equivalence:

$$
\Sigma_{s,t}^{\infty} V/W \longrightarrow \Sigma_{s,t}^{\infty} X/U .
$$

Proof. By Remark 2.17 and Lemma 2.19, the pushout square of representable spaces associated to an upper distinguished square is a homotopy pushout square. This implies item (a) because the suspension functor $\Sigma_{s,t}^{\infty}$ preserves homotopy pushout squares.

To prove (b), we use 2.19 to conclude there is even an isomorphism between the underlying pointed spaces. □

In the next result, the projective line $\mathbb{P}^1$ is pointed by the rational point at infinity. Another piece of useful information about mixed spheres is:

Lemma 2.21 *In* $\mathrm{SH}(k)$*, there are canonical isomorphisms:*

(a) $\Sigma_{s,t}^{\infty} (\mathbb{A}^1/\mathbb{A}^1 \setminus 0) = \Sigma_{s,t}^{\infty} (S_s^1 \wedge S_t^1)$.

(b) $\Sigma_{s,t}^{\infty} (\mathbb{P}^1, \infty) = \Sigma_{s,t}^{\infty} (S_s^1 \wedge S_t^1)$.

Proof. To prove (a), we use 2.12, and 2.16 to conclude there is a distinguished triangle:

$$\Sigma_{s,t}^{\infty} (\mathbb{A}^1 \setminus 0, 1) \longrightarrow \Sigma_{s,t}^{\infty} (\mathbb{A}^1, 1) \longrightarrow \Sigma_{s,t}^{\infty} (\mathbb{A}^1/\mathbb{A}^1 \setminus 0) \longrightarrow \Sigma_{s,t}^{\infty} (\mathbb{A}^1 \setminus 0, 1)[1] \ .$$

To prove (b), cover the projective line by two affine lines; the choice of a point on $\mathbb{P}^1$ is not important, since all such points are $\mathbb{A}^1$-homotopic. Moreover, by 2.20(a), there is a distinguished triangle

$$\Sigma_{s,t}^{\infty} (\mathbb{A}^1 \setminus 0)_+ \longrightarrow \Sigma_{s,t}^{\infty} \mathbb{A}_+^1 \vee \Sigma_{s,t}^{\infty} \mathbb{A}_+^1 \longrightarrow \Sigma_{s,t}^{\infty} \mathbb{P}_+^1 \longrightarrow \Sigma_{s,t}^{\infty} (\mathbb{A}^1 \setminus 0)_+[1] \ .$$

To remove the disjoint base-points, we point all spaces by $1 \colon \mathrm{Spec}(k) \to \mathbb{A}^1 \setminus 0$ and consider the quotients. For instance, we have

$$\Sigma_{s,t}^{\infty}(\mathbb{A}^1 \setminus 0)_+/\Sigma_{s,t}^{\infty}\mathrm{Spec}(k)_+ = \Sigma_{s,t}^{\infty}(\mathbb{A}^1 \setminus 0, 1) \ .$$

By considering the resulting distinguished triangle, the claim follows from homotopy invariance 2.12. □

Remark 2.22 *The pointed space* $T \colon = \mathbb{A}^1/\mathbb{A}^1 \setminus 0$ *is called the Tate sphere. There is also a 'motivic unstable homotopy category' for spaces, and 2.21 holds unstably. We refer the reader to Sect. 5.2 for the unstable theory.*

We place emphasis on a typical instance of 2.20(b):

Corollary 2.23 *Suppose there is an etale morphism*

$$p \colon V \longrightarrow X \ ,$$

in Sm/k*, and* Z *is a closed sub-scheme of* X *such that there is an isomorphism*

$$p^{-1}(Z) \longrightarrow Z \ .$$

Then there is a canonical isomorphism in $\mathrm{SH}(k)$*:*

$$\Sigma_{s,t}^{\infty} (V/V \setminus p^{-1}(Z)) = \Sigma_{s,t}^{\infty} (X/X \setminus Z) \ .$$

As in classical homotopy theory, there is also a notion of Thom spaces in motivic homotopy theory.

Definition 2.24 *Suppose there is a vector bundle $\mathcal{E} \to X$ in Sm/k, with zero section $i\colon X \to \mathcal{E}$. Its Thom space is defined by setting*

$$\mathrm{Th}(\mathcal{E}/X)\colon = \mathcal{E}/\mathcal{E} \smallsetminus i(X) \,,$$

pointed by the image of $\mathcal{E} \smallsetminus i(X)$.

Example 2.25 *We give some examples of Thom spaces.*

(a) Suppose we have given vector bundles

$$\mathcal{E}_1 \longrightarrow X_1, \mathcal{E}_2 \longrightarrow X_2$$

in Sm/k. Then, as pointed spaces

$$\mathrm{Th}(\mathcal{E}_1 \times \mathcal{E}_2/X_1 \times X_2) = \mathrm{Th}(\mathcal{E}_1/X_1) \wedge \mathrm{Th}(\mathcal{E}_2/X_2) \,.$$

(b) The Thom space of the trivial 1-bundle

$$\mathbb{A}^1 \times X \longrightarrow X$$

is the smash product $T \wedge X_+$.

Theorem 2.26 (Homotopy Purity) *Suppose there is a closed embedding in Sm/k*

$$i\colon Z \lhook\joinrel\longrightarrow X \,.$$

Denote the corresponding normal vector bundle of Z in X by $\mathbb{N}_{X,Z}$.
Then there is a canonical isomorphism in $\mathrm{SH}(k)$:

$$\Sigma^\infty_{s,t} \mathrm{Th}(\mathbb{N}_{X,Z}) = \Sigma^\infty_{s,t} (X/X \smallsetminus i(Z)) \,.$$

To prove the general case of the homotopy purity theorem, one employs the well-known deformation to the normal cone construction based on the blow-up of $X \times \mathbb{A}^1$ with center in $Z \times \{0\}$. This is an algebraic analog of the notion of tubular neighborhood in topology. In what follows, by 'isomorphism' we mean a canonical isomorphism in $\mathrm{SH}(k)$.

Next we construct the isomorphism in the homotopy purity theorem for any finite separable field extension k'/k. By the Primitive Element Theorem, there exists an element α such that $k' = k(\alpha)$. Consider the surjective map

$$\phi\colon k[X] \longrightarrow k', X \longmapsto \alpha \,.$$

If f is the minimal polynomial of α, then the induced closed embedding

$$i\colon \mathrm{Spec}(k') \lhook\joinrel\longrightarrow \mathbb{A}^1_k$$

sends the closed point of $\mathrm{Spec}(k')$ to the closed point x of $\mathbb{A}^1_k$, corresponding to the prime ideal (f). We claim there is an isomorphism

$$\Sigma_{s,t}^{\infty} \, \mathrm{Th}(\mathbb{N}_{\mathbb{A}_k^1, k'}) = \Sigma_{s,t}^{\infty} \, (\mathbb{A}_k^1/\mathbb{A}_k^1 \smallsetminus \{x\}) \,.$$

With respect to the identification of the normal bundle of i with $\mathbb{A}_{k'}^1$, the Thom space $\mathrm{Th}(\mathbb{N}_{\mathbb{A}_k^1, k'})$ is isomorphic to $\mathbb{A}_{k'}^1/\mathbb{A}_{k'}^1 \smallsetminus \{0\}$.

In $k'[X]$, f factors into irreducible polynomials, say $f_1, \ldots, f_q$. Denote by $x_1, \ldots, x_q$ the corresponding closed points. Since the extension k'/k is separable, we may assume $f_1 = X - \alpha$ and $f_i(\alpha) \neq 0$ if $i \neq 1$. We have an automorphism

$$k'[X] \longrightarrow k'[X], X \longmapsto X - \alpha \,.$$

By the above, there is an isomorphism

$$\Sigma_{s,t}^{\infty} \, \mathrm{Th}(\mathbb{N}_{\mathbb{A}_k^1, k'}) = \Sigma_{s,t}^{\infty} \, (\mathbb{A}_{k'}^1/\mathbb{A}_{k'}^1 \smallsetminus \{x_1\}) \,.$$

We note that

$$\mathbb{A}_{k'}^1 \smallsetminus \{x_1\} \lhook\joinrel\longrightarrow \mathbb{A}_{k'}^1$$

fits into the upper distinguished square:

$$
\begin{array}{ccc}
\mathbb{A}_{k'}^1 \smallsetminus \{x_1, x_2, \ldots, x_q\} & \longrightarrow & \mathbb{A}_{k'}^1 \smallsetminus \{x_2, \ldots, x_q\} \\
\downarrow & & \downarrow \\
\mathbb{A}_{k'}^1 \smallsetminus \{x_1\} & \longrightarrow & \mathbb{A}_{k'}^1
\end{array}
$$

By applying 2.20(b), we find

$$\Sigma_{s,t}^{\infty} \, \mathrm{Th}(\mathbb{N}_{\mathbb{A}_k^1, k'}) = \Sigma_{s,t}^{\infty} \, (\mathbb{A}_{k'}^1 \smallsetminus \{x_2, \ldots, x_q\}/\mathbb{A}_{k'}^1 \smallsetminus \{x_1, x_2, \ldots, x_q\}) \,.$$

The right hand side of this isomorphism is related to $\mathbb{A}_k^1/\mathbb{A}_k^1 \smallsetminus \{x\}$ via the upper distinguished square

$$
\begin{array}{ccc}
\mathbb{A}_{k'}^1 \smallsetminus \{x_1, x_2, \ldots, x_q\} & \longrightarrow & \mathbb{A}_{k'}^1 \smallsetminus \{x_2, \ldots, x_q\} \\
\downarrow & & \downarrow \\
\mathbb{A}_k^1 \smallsetminus \{x\} & \longrightarrow & \mathbb{A}_k^1
\end{array}
$$

which defines an isomorphism

$$\Sigma_{s,t}^{\infty} \, (\mathbb{A}_{k'}^1 \smallsetminus \{x_2, \ldots, x_q\}/\mathbb{A}_{k'}^1 \smallsetminus \{x_1, x_2, \ldots, x_q\}) = \Sigma_{s,t}^{\infty} \, (\mathbb{A}_k^1/\mathbb{A}_k^1 \smallsetminus \{x\}) \,.$$

By using the fact that the isomorphism of 2.26 is compatible in the obvious sense with etale morphisms, one shows that the isomorphism we constructed above coincides with the isomorphism in the homotopy purity theorem.

Remark 2.27 *Given a finite etale map*

$$Y \longrightarrow X \,,$$

the proof of 2.26 implies there exists, in $\mathrm{SH}(k)$, *a transfer map*

$$\Sigma_{s,t}^{\infty} \, X_+ \longrightarrow \Sigma_{s,t}^{\infty} \, Y_+ \,.$$

3 Cohomology Theories

The introduction of the motivic stable homotopy category via spectra provides a convenient framework for defining cohomology theories of algebraic varieties. Such theories encode important data about the input, often in a form which allows to make algebraic manipulations. We shall consider the examples of motivic cohomology, algebraic K-theory and algebraic cobordism.

3.1 The Motivic Eilenberg-MacLane Spectrum H$\mathbb{Z}$

In stable homotopy theory, the key to understand cohomology theories is to study their representing spectra. Singular cohomology is a prime example. This motivates the construction of the motivic Eilenberg-MacLane spectrum, which we denote by **H$\mathbb{Z}$**.

Let $K(\mathbb{Z}, n)$ denote the Eilenberg-MacLane simplicial set with homotopy groups

$$\pi_i K(\mathbb{Z}, n) = \begin{cases} \mathbb{Z} & i = n, \\ 0 & i \neq n. \end{cases} \tag{1}$$

The simplicial set $K(\mathbb{Z}, n)$ is uniquely determined up to weak equivalence by (1). A model for $K(\mathbb{Z}, n)$ can be constructed as follows: Denote by $\mathbb{Z}(-)$ the functor which associates to a pointed simplicial set (X, x) the simplicial abelian group $\mathbb{Z}[X]/\mathbb{Z}[x]$. That is, the simplicial free abelian group generated by X modulo the copy of the integers generated by the basepoint. Then $K(\mathbb{Z}, n)$ is the underlying simplicial set of the simplicial abelian group $\mathbb{Z}(\Delta[n]/\partial\Delta[n])$. Moreover, the $K(\mathbb{Z}, n)$'s assemble to define a spectrum $H\mathbb{Z}$ which represents singular cohomology: The integral singular cohomology group $H^n(X, \mathbb{Z})$ coincides with $\mathrm{Hom}_{\mathrm{SH}}(\Sigma^\infty X_+, H\mathbb{Z})$.

In motivic homotopy theory, there is a closely related algebro-geometric analog of the spectrum $H\mathbb{Z}$. But as it turns out, it is impossible to construct an (s, t)-bispectrum whose constituent terms satisfy a direct motivic analog of (1); however, by using the theory of algebraic cycles one can define an (s, t)-spectrum **H$\mathbb{Z}$** that represents motivic cohomology. Next, we will indicate the construction of **H$\mathbb{Z}$**.

If $X \in \mathrm{Sm}/k$, let $L(X)$ be the following functor: Its value on $U \in \mathrm{Sm}/k$ is the free abelian group generated by closed irreducible subsets of $U \times X$ which are finite over U and surjective over a connected component of U. For simplicity, we refer to elements in this group as cycles. The graph $\Gamma(f)$ of a morphism

$$f \colon U \longrightarrow X$$

is an example of a cycle in $U \times X$.

It turns out that $L(X)$ is a Nisnevich sheaf, hence a pointed space by forgetting the abelian group structure. Moreover, there is a map

$$\Gamma(X) \colon X \longrightarrow L(X) \,.$$

One can extend $L(-)$ to a functor from pointed spaces to Nisnevich sheaves with values in simplicial abelian groups.[5] For example, we have that

$$L(S_s^0 \wedge S_t^1) = L((\mathbb{A}^1 \setminus 0, 1))$$

is the quotient sheaf of abelian groups $L(\mathbb{A}^1 \setminus 0)/L(\mathrm{Spec}(k))$, considered as a pointed space. The pointed space $L(S_s^1 \wedge S_t^1)$ turns out to be equivalent to

$$L(\mathbb{P}^1, \infty) = L(\mathbb{P}^1)/L(\mathrm{Spec}(k)) \,.$$

Remark 3.1 *One can show that $L(\mathbb{P}^1, \infty)$ is weakly equivalent to the infinite projective space $\mathbb{P}^\infty$. Hence, if k admits a complex embedding, taking complex points yields an equivalence*

$$L(\mathbb{P}^1, \infty)(\mathbb{C}) \sim \mathbb{CP}^\infty = K(\mathbb{Z}, 2).$$

The exterior product of cycles induces a pairing

$$L(S_s^p \wedge S_t^q) \wedge L(S_s^m \wedge S_t^n) \longrightarrow L(S_s^{p+m} \wedge S_t^{q+n}) \,.$$

In particular, we obtain the composite maps

$$\sigma_s \colon S_s^1 \wedge L(S_s^m \wedge S_t^n) \xrightarrow{\Gamma(S_s^1) \wedge \mathrm{id}} L(S_s^1) \wedge L(S_s^m \wedge S_t^n) \longrightarrow L(S_s^{m+1}, S_t^n) \,,$$

$$\sigma_t \colon S_t^1 \wedge L(S_s^m \wedge S_t^n) \xrightarrow{\Gamma(S_t^1) \wedge \mathrm{id}} L(S_t^1) \wedge L(S_s^m \wedge S_t^n) \longrightarrow L(S_s^m, S_t^{n+1}) \,.$$

$$(2)$$

Remark 3.2 *A topologically inclined reader might find it amusing to compare the above with Bökstedt's notion of functors with smash products.*

Definition 3.3 *The Eilenberg-MacLane spectrum $\mathbf{H}\mathbb{Z}$ is the (s, t)-bispectrum with constituent pointed spaces $\mathbf{H}\mathbb{Z}_{m,n} \colon = L(S_s^m \wedge S_t^n)$ and structure maps given by (2).*

Remark 3.4 *Let ℓ be a prime number. The Eilenberg-MacLane spectrum with mod-ℓ coefficients is defined as above by taking the reduction of $L(X)$ modulo ℓ in the category of abelian sheaves.*

We can now define the motivic cohomology groups of (s, t)-bispectra.

Definition 3.5 *Let E be an (s, t)-bispectrum and let p, q be integers. The integral motivic cohomology groups of E are defined by*

$$\mathbf{H}\mathbb{Z}^{p,q}(E) \colon = \mathrm{Hom}_{\mathrm{SH}(k)}(E, S_s^{p-q} \wedge S_t^q \wedge \mathbf{H}\mathbb{Z}) \,.$$

---

[5] Since every pointed space is a colimit of representable functors and L preserves colimits, it suffices to describe the values of L on Sm/k.

Suppose $X \in \mathrm{Sm}/k$. The integral motivic cohomology group $\mathbf{H}^{p,q}(X, \mathbb{Z})$ of X in degree p and weight q is by definition $\mathbf{H}\mathbb{Z}^{p,q}(\Sigma_{s,t}^{\infty} X_+)$. One can show that $\mathbf{H}^{p,q}(X, \mathbb{Z})$ is isomorphic to the higher Chow group $\mathrm{CH}^q(X, 2q - p)$ introduced by Bloch in [Blo86].

In Sect. 4, we shall outline an approach to construct a spectral sequence whose E_2-terms are the integral motivic cohomology groups of X. Its target groups are the algebraic K-groups of X. Since there is a spectral sequence for topological K-theory whose input terms are singular cohomology groups, this would allow to make more precise the analogy between motivic and singular cohomology.

3.2 The Algebraic K-Theory Spectrum KGL

In 2.21(b), we noted that the suspension (s, t)-bispectra of $(\mathbb{P}^1, \infty)$ and $S_s^1 \wedge S_t^1$ are canonically isomorphic in $\mathrm{SH}(k)$. In fact, we may replace the suspension coordinates S_s^1 and S_t^1 by $\mathbb{P}^1$ without introducing changes in the motivic stable homotopy category. In order to represent cohomology theories on Sm/k, it turns out to be convenient to consider $\mathbb{P}^1$-spectra. To define the category of such spectra, one replaces in 2.1 every occurrence of the simplicial circle by the projective line. If E is a $\mathbb{P}^1$-spectrum, we may associate a bigraded cohomology theory by setting

$$E^{p,q}(X) := \mathrm{Hom}_{\mathrm{SH}(k)}(\Sigma_{\mathbb{P}^1}^{\infty} X_+, E \wedge S_s^{p-2q} \wedge (\mathbb{P}^1)^{\wedge q}) . \tag{3}$$

An important example is the spectrum representing algebraic K-theory, which we describe next.

If m is a non-negative integer, we denote the Grassmannian of vector spaces of dimension n in the $n + m$-dimensional vector space over k by $\mathrm{Gr}_n(\mathbb{A}^{n+m})$. By letting m tend to infinity, it results a directed system of spaces; denote the colimit by BGL_n. There are canonical monomorphisms

$$\cdots \overset{\subset}{\longrightarrow} \mathrm{BGL}_n \overset{\subset}{\longrightarrow} \mathrm{BGL}_{n+1} \overset{\subset}{\longrightarrow} \cdots .$$

Denote by BGL the sequential colimit of this diagram.

Let KGL be a fibrant replacement of

$$\mathbb{Z} \times \mathrm{BGL}$$

in the unstable motivic homotopy theory of k. Then there exists a map

$$\beta \colon \mathbb{P}^1 \wedge \mathrm{KGL} \longrightarrow \mathrm{KGL} , \tag{4}$$

which represents the canonical Bott element of

$$K_0(\mathbb{P}^1 \wedge (\mathbb{Z} \times \mathrm{BGL})) .$$

More precisely, the map (4) is adjoint to a lift of the isomorphism

$$\mathbb{Z} \times \mathrm{BGL} \longrightarrow \Omega_{\mathbb{P}^1}(\mathbb{Z} \times \mathrm{BGL})$$

in the unstable motivic homotopy category which induces Bott periodicity in algebraic K-theory. Details are recorded in [Voe98].

Definition 3.6 *The algebraic K-theory spectrum* **KGL** *is the $\mathbb{P}^1$-spectrum*

$$(\mathrm{KGL}, \mathrm{KGL}, \cdots, \mathrm{KGL}, \dots),$$

together with the structure maps in (4).

As in topological K-theory, there is also Bott periodicity in algebraic K-theory: The structure maps of **KGL** are defined by lifting an isomorphism in the unstable motivic homotopy category, so there is an isomorphism

$$\mathbb{P}^1 \wedge \mathbf{KGL} = \mathbf{KGL} . \tag{5}$$

If k admits a complex embedding, taking $\mathbb{C}$-points defines a realization functor

$$t_{\mathbb{C}} \colon \mathrm{SH}(k) \longrightarrow \mathrm{SH} .$$

This functor sends $\Sigma_{s,t}^{\infty} S^{p,q}$ to the suspension spectrum of the p-sphere and **KGL** to the ordinary complex topological K-theory spectrum.

3.3 The Algebraic Cobordism Spectrum MGL

In what follows, we use the notation in 3.2. Denote the tautological vector bundle over the Grassmannian by

$$\gamma_{n,m} \longrightarrow \mathrm{Gr}_n(\mathbb{A}^{n+m})$$

The canonical morphism

$$\mathrm{Gr}_n(\mathbb{A}^{n+m}) \longrightarrow \mathrm{Gr}_n(\mathbb{A}^{n+m+1})$$

is covered by a bundle map $\gamma_{n,m} \longrightarrow \gamma_{n,m+1}$. Taking the colimit over m yields the universal n-dimensional vector bundle

$$\gamma_n \longrightarrow \mathrm{BGL}_n .$$

The product

$$\mathbb{A}^1 \times \gamma_n \longrightarrow \mathrm{BGL}_n$$

with the trivial one-dimensional bundle

$$\mathbb{A}^1 \longrightarrow \mathrm{Spec}(k)$$

is classified by the canonical map

$$\mathrm{BGL}_n \longrightarrow \mathrm{BGL}_{n+1} \, .$$

In particular, there exists a bundle map

$$\mathbb{A}^1 \times \gamma_n \longrightarrow \gamma_{n+1} \, .$$

On the level of Thom spaces, we obtain the map

$$\mathrm{Th}(\mathbb{A}^1) \wedge \mathrm{Th}(\gamma_n) = (\mathbb{A}^1/\mathbb{A}^1 \smallsetminus 0) \wedge \mathrm{Th}(\gamma_n) \longrightarrow \mathrm{Th}(\gamma_{n+1}) \, . \tag{6}$$

Here, note that we may apply 2.25(a) since the map between γ_n and BGL_n is a colimit of vector bundles of smooth schemes. From (6) and Remark 2.22, we get

$$c_n \colon \mathbb{P}^1 \wedge \mathrm{Th}(\gamma_n) \longrightarrow \mathrm{Th}(\gamma_{n+1}) \, . \tag{7}$$

Definition 3.7 *The algebraic cobordism spectrum* **MGL** *is the* $\mathbb{P}^1$-*spectrum*

$$(\mathrm{Th}(\gamma_0), \mathrm{Th}(\gamma_1), \dots, \mathrm{Th}(\gamma_n), \dots) \, ,$$

together with the structure maps in (7).

The algebraic cobordism spectrum is the motivic analog of the ordinary complex cobordism spectrum MU. One can check that

$$t_{\mathbb{C}}(\mathbf{MGL}) = \mathrm{MU} \, .$$

The notions of orientation, and formal group laws in ordinary stable homotopy theory have direct analogs for $\mathbb{P}^1$-spectra.

4 The Slice Filtration

In classical homotopy theory, the Eilenberg-MacLane space $K(\mathbb{Z}, n)$ has a unique non-trivial homotopy group. And up to homotopy equivalence there is a unique such space for each n (1). The situation in motivic homotopy theory is quite different. For example, the homotopy groups $\pi_{p,q}(\mathbf{H}\mathbb{Z})$ are often non-zero, as one may deduce from the isomorphism between $\pi_{q,q}(\mathbf{H}\mathbb{Z})$ and the Milnor K-theory of k. To give an internal description of $\mathbf{H}\mathbb{Z}$ within the stable motivic homotopy category, we employ the so-called slice filtration. In what follows, we recall and discuss the status of Voevodsky's conjectures about the slices of the sphere spectrum $\mathbf{1} = \Sigma_{s,t}^\infty \mathrm{Spec}(k)_+$, $\mathbf{H}\mathbb{Z}$, and $\mathbf{KGL}$. For more details, we refer to the original papers [Voe02b] and [Voe02c].

Let $\mathrm{SH}^{\mathrm{eff}}(k)$ denote the smallest triangulated sub-category of $\mathrm{SH}(k)$ which is closed under direct sums and contains all (s, t)-bispectra of the form $\Sigma_{s,t}^\infty X_+$. If $n \geq 1$, the desuspension spectrum $\Sigma_t^{-n} \Sigma_{s,t}^\infty X_+$ is not contained in $\mathrm{SH}^{\mathrm{eff}}(k)$.

The 'effective' s-stable homotopy category $\mathrm{SH}_s^{\mathrm{eff}}(k)$ is defined similarly, by replacing $\Sigma_{s,t}^\infty$ with the s-suspension Σ_s^∞.

In this section, we shall study the sequence of full embeddings of categories

$$\cdots \hookrightarrow \Sigma_{s,t}^1 \mathrm{SH}^{\mathrm{eff}}(k) \hookrightarrow \mathrm{SH}^{\mathrm{eff}}(k) \hookrightarrow \Sigma_{s,t}^{-1}\mathrm{SH}^{\mathrm{eff}}(k) \hookrightarrow \cdots . \qquad (8)$$

The sequence (8) is called the slice filtration. For an alternative formulation of the slice filtration, consider [Lev03]. The above is a filtration in the sense that $\mathrm{SH}(k)$ is the smallest triangulated category which contains $\Sigma_{s,t}^n \mathrm{SH}^{\mathrm{eff}}(k)$ for all n and is closed under arbitrary direct sums in $\mathrm{SH}(k)$. For each n, the category $\Sigma_{s,t}^n \mathrm{SH}^{\mathrm{eff}}(k)$ is triangulated. Moreover, this category has arbitrary direct sums and a set of compact generators.

Neeman's work on triangulated categories in [Nee96] shows that the full inclusion functor

$$i_n : \Sigma_{s,t}^n \mathrm{SH}^{\mathrm{eff}}(k) \hookrightarrow \mathrm{SH}(k)$$

has a right adjoint

$$r_n : \mathrm{SH}(k) \longrightarrow \Sigma_{s,t}^n \mathrm{SH}^{\mathrm{eff}}(k) .$$

such that the unit of the adjunction is an isomorphism

$$\mathrm{Id} \longrightarrow r_n \circ i_n .$$

Consider now the reverse composition

$$f_n : \mathrm{SH}(k) \xrightarrow{\ r_n\ } \Sigma_{s,t}^n \mathrm{SH}^{\mathrm{eff}}(k) \xrightarrow{\ i_n\ } \mathrm{SH}(k) .$$

The counit

$$f_{n+1} \longrightarrow \mathrm{Id}$$

applied to the functor f_n determines a natural transformation

$$f_{n+1} = f_{n+1} \circ f_n \longrightarrow f_n .$$

If E is an (s,t)-bispectrum and $n \in \mathbb{Z}$, then the slice tower of E consists of the distinguished triangles in $\mathrm{SH}(k)$

$$f_{n+1}E \longrightarrow f_n E \longrightarrow s_n E \longrightarrow f_{n+1}E[1] . \qquad (9)$$

Definition 4.1 *The n-th slice of E is $s_n E$.*

Remark 4.2 *Since $f_{n+1}E, f_n E \in \Sigma_{s,t}^n \mathrm{SH}^{\mathrm{eff}}(k)$, we get that $s_n E \in \Sigma_{s,t}^n$ $\mathrm{SH}^{\mathrm{eff}}(k)$. The adjunction above implies that $s_n E$ receives only the trivial map from $\Sigma_{s,t}^{n+1}\mathrm{SH}^{\mathrm{eff}}(k)$. Standard arguments show that these properties character-ize, up to canonical isomorphism, the triangulated functors*

$$s_n \colon \mathrm{SH}(k) \longrightarrow \mathrm{SH}(k) \ .$$

Suppose that $E \in \Sigma^n_{s,t}\mathrm{SH}^{\mathrm{eff}}(k)$ and let $k \leq n$. Then $f_k E = E$ and the k-th slice $s_k E$ of E is trivial for all $k < n$.

Slice towers are analogous to Postnikov towers in algebraic topology; the slices corresponding to the cofibers or quotients. If E is an ordinary spectrum, recall that its Postnikov tower expresses E as the sequential colimit of a diagram of cofibrations

$$\cdots \longrightarrow P_{-1}E \xrightarrow{p_0} P_0 E \xrightarrow{p_1} P_1 E \longrightarrow \cdots \xrightarrow{p_n} P_n E \xrightarrow{p_{n+1}} \cdots \ .$$

The canonical map

$$P_n E \longrightarrow E$$

induces isomorphisms on all stable homotopy groups π_i and $i \leq n$, whereas for $i > n$ we have

$$\pi_i(P_n E) = 0 \ .$$

It follows that the cofiber of p_n is an Eilenberg-MacLane spectrum $\Sigma^n H\pi_n(E)$; hence, in particular an $H\mathbb{Z}$-module. Now recall that the zeroth stage of the Postnikov tower of the ordinary sphere spectrum is the Eilenberg-MacLane spectrum $H\mathbb{Z}$.

This gives some topological motivation for the following conjecture, which in turn implies a characterization of motivic cohomology entirely in terms of the motivic stable homotopy category.

Conjecture 1. $s_0 \mathbf{1} = \mathbf{H}\mathbb{Z}$.

The collection of the functors s_n is compatible with the smash product, meaning that if E and F are objects of $\mathrm{SH}(k)$, there is a natural map

$$s_n(E) \wedge s_m(F) \longrightarrow s_{n+m}(E \wedge F) \ .$$

In particular, we get a map

$$s_0(\mathbf{1}) \wedge s_n(E) \longrightarrow s_n(E) \ .$$

If Conjecture 1 holds, then the above allows us to conclude that each slice of an (s,t)-bispectrum has a natural and unique structure of a module over the motivic cohomology spectrum.

Theorem 4.3 *Conjecture 1 holds for all fields of characteristic zero.*

Remark 4.4 *According to [Lev03, Theorem 8.4.1], Theorem 4.5 holds for every perfect field.*

In our sketch proof of 4.3, we start with the zeroth slice of $\mathbf{HZ}$.

Lemma 4.5 *Let k be a field of characteristic zero. Then $s_0\mathbf{HZ} = \mathbf{HZ}$.*

To prove 4.5, we will make use of the following facts: First, the motivic Eilenberg-MacLane spectrum is an effective spectrum. Thus

$$f_0\mathbf{HZ} = \mathbf{HZ} \,.$$

Second, if X is a scheme in Sm/k, then every map in $\mathrm{SH}(k)$ from $\Sigma^1_{s,t}\Sigma^\infty_{s,t} X_+$ to $\mathbf{HZ}$ is trivial by Voevodsky's cancellation theorem [Voe02a].[6]

Hence, for every $E \in \Sigma^1_{s,t}\mathrm{SH}^{\mathrm{eff}}(k)$, we get isomorphisms

$$\mathrm{Hom}_{\Sigma^1_{s,t}\mathrm{SH}^{\mathrm{eff}}(k)}(E, r_1\mathbf{HZ}) = \mathrm{Hom}_{\mathrm{SH}(k)}(i_1E, \mathbf{HZ})$$
$$= 0 \,.$$

This proves

$$r_1\mathbf{HZ} = 0 \,.$$

In particular, $f_1\mathbf{HZ} = 0$ and hence

$$s_0\mathbf{HZ} = f_0\mathbf{HZ}$$
$$= \mathbf{HZ} \,.$$

Denote by $\mathcal{C}$ the cone of the unit map

$$1 \longrightarrow \mathbf{HZ} \,.$$

To finish the proof of 4.3, note that by 4.5 it remains to show

$$s_0\mathcal{C} = 0 \,.$$

Remark 4.2 shows it suffices to prove that $\mathcal{C}$ is contained in $\Sigma^1_{s,t}\mathrm{SH}^{\mathrm{eff}}(k)$.

If k is a field of characteristic zero, then $\mathbf{HZ}$ has an explicit description in terms of infinite symmetric products of $\mathbb{A}^n$ and $\mathbb{A}^n \smallsetminus 0$. This allows to conclude the statement about $\mathcal{C}$. We shall sketch a proof.

Recall that

$$\mathbf{HZ}_{n,n} = L(S^n_s \wedge S^n_t)$$

is weakly equivalent to the quotient sheaf $L(\mathbb{A}^n)/L(\mathbb{A}^n \smallsetminus 0)$.

Let $L^{\mathrm{eff}}(\mathbb{A}^n)$ denote the sheaf which maps $U \in \mathrm{Sm}/k$ to the free abelian monoid generated by closed irreducible subsets of $U \times \mathbb{A}^n$ which are finite over U and also surjective over a connected component of U.

The sheaf $L^{\mathrm{eff}}(\mathbb{A}^n \smallsetminus 0)$ is defined similarly. In particular, L^{eff} consists of cycles with nonnegative coefficients. Denote the quotient $L^{\mathrm{eff}}(\mathbb{A}^n)/L^{\mathrm{eff}}(\mathbb{A}^n \smallsetminus 0)$

---

[6] In other words, motivic cohomology of X in weight -1 is zero.

by $\mathbf{HZ}_{n,n}^{\text{eff}}$. It is straightforward to define $\mathbf{HZ}_{m,n}^{\text{eff}}$ and moreover note that, along the same lines as above, these pointed spaces form an (s,t)-bispectrum $\mathbf{HZ}^{\text{eff}}$.

If $n \geq 1$, the canonical map

$$\mathbf{HZ}_{n,n}^{\text{eff}} \longrightarrow \mathbf{HZ}_{n,n}$$

turns out to be a weak equivalence of pointed spaces.

Furthermore, the canonical inclusion

$$1 \longrightarrow \mathbf{HZ}$$

factors through

$$1 \longrightarrow \mathbf{HZ}^{\text{eff}}.$$

Hence the quotient of the latter map is equivalent to $\mathcal{C}$.

Next, we consider $T^n = \mathbb{A}^n/\mathbb{A}^n \smallsetminus 0$. Let $\mathbf{HZ}_{\bar{n},n}^{\leq d}$ be the subsheaf of $\mathbf{HZ}_{n,n}^{\text{eff}}$ mapping $U \in \text{Sm}/k$ to the cycles of degree $\leq d$ over U. Then the natural inclusion

$$T^n \hookrightarrow \mathbf{HZ}_{n,n}^{\text{eff}}$$

has a filtration

$$T^n = \mathbf{HZ}_{\bar{n},n}^{\leq 1} \hookrightarrow \mathbf{HZ}_{\bar{n},n}^{\leq 2} \hookrightarrow \cdots \hookrightarrow \mathbf{HZ}_{\bar{n},n}^{\leq d} \hookrightarrow \cdots \hookrightarrow \mathbf{HZ}_{n,n}^{\text{eff}}.$$

Taking quotients $\mathbf{HZ}_{\bar{n},n}^{\leq d}/\mathbf{HZ}_{\bar{n},n}^{\leq d-1}$ induces a filtration of the quotient sheaf $\mathbf{HZ}_{n,n}^{\text{eff}}/T^n$. Results of Suslin-Voevodsky [SV96] imply that $\mathbf{HZ}_{\bar{n},n}^{\leq d}/\mathbf{HZ}_{\bar{n},n}^{\leq d-1}$ is isomorphic to the d-th symmetric power sheaf $(T^n)^{\wedge d}/S_d$ of T^n, where S_d denotes the symmetric group on d letters. This uses the characteristic zero assumption.

One can show that $(T^n)^{\wedge d}/S^d$ is contained in the smallest class of pointed spaces which is closed under homotopy colimits and contains all $X/(X - Z)$, where Z is a closed subscheme of $X \in \text{Sm}/S$ of codimension $\geq n+1$. After resolving all singularities in Z, homotopy purity 2.26 implies that the (s,t)-suspension spectrum of any space in this class is contained in $\Sigma_{s,t}^{n+1}\text{SH}^{\text{eff}}(k)$. This ends our sketch proof of 4.3.

Analogous to (8), there exists a slice filtration in $\text{SH}_s^{\mathbb{A}^1}(k)$. For $n \geq 0$, let $\Sigma_t^n \text{SH}_s^{\mathbb{A}^1}(k)$ be the smallest compactly generated triangulated sub-category of $\text{SH}_s^{\mathbb{A}^1}(k)$ which is closed under arbitrary direct sums, and generated by the objects $\Sigma_t^n \Sigma_s^\infty X_+$. We obtain the filtration

$$\cdots \hookrightarrow \Sigma_t^n \text{SH}_s^{\mathbb{A}^1}(k) \longrightarrow \Sigma_t^{n-1}\text{SH}_s^{\mathbb{A}^1}(k) \cdots \hookrightarrow \Sigma_t^0 \text{SH}_s^{\mathbb{A}^1}(k) = \text{SH}_s^{\mathbb{A}^1}(k) \ .$$

The t-suspension functor Σ_t^∞ preserves the slice filtrations in the sense that

$$\Sigma_t^\infty(\Sigma_t^n \text{SH}_s^{\mathbb{A}^1}(k)) \subseteq \Sigma_{s,t}^n \text{SH}^{\text{eff}}(k) \ .$$

Denote the right adjoint of this inclusion by

$$\Omega_t^\infty : \mathrm{SH}(k) \longrightarrow \mathrm{SH}_s^{\mathbb{A}^1}(k) .$$

Conjecture 2. There is an inclusion $\Omega_t^\infty(\Sigma_{s,t}^n\mathrm{SH}^{\mathrm{eff}}(k)) \subseteq \Sigma_t^n\mathrm{SH}_s^{\mathbb{A}^1}(k)$.

In topology, recall that first applying the suspension functor, and second the loop space functor preserve connectivity. An inductive argument shows that the following implies Conjecture 2.

Conjecture 3. If $X \in \mathrm{Sm}/k$ and $n \geq 0$, then $\Omega_t^1\Sigma_t^1\Sigma_s^\infty(S_t^n \wedge X_+) \in \Sigma_t^n\mathrm{SH}_s^{\mathbb{A}^1}(k)$.

This conjecture is not known at present, even if k is a field of characteristic zero. A proof seems to require a fair amount of work. A possible approach to prove Conjecture 3 is to develop an analog of the theory of operads, at least for A_∞-operads, in order to have an explicit model for $\Omega_{\mathbb{P}^1}\Sigma_{\mathbb{P}^1}(X)$. Framed correspondences might be a useful tool in working out such a theory.

At last in the section, we relate the above machinery to a possibly new approach to some recent advances in algebraic K-theory.

In [Bei87], Beilinson conjectured the existence of an Atiyah-Hirzebruch type spectral sequence for the algebraic K-groups of nice schemes. In [BL95], Bloch and Lichtenbaum constructed such a spectral sequence. Their work has been expanded by Friedlander and Suslin [FS02], by Levine [Lev01], and by Suslin [Sus03].

The slice tower (9) acquires an associated spectral sequence. In the example of **KGL**, one obtains

$$\mathrm{Hom}_{\mathrm{SH}(k)}(\Sigma_{s,t}^\infty X_+, S_s^{p-q} \wedge S_t^q \wedge s_0\mathbf{KGL}) \Rightarrow K_{-p-q}(X) . \tag{10}$$

The main problem with this spectral sequence is to identify the input terms in (10) with motivic cohomology.

Conjecture 4. $s_0\mathbf{KGL} = \mathbf{H}\mathbb{Z}$.

Because of Bott periodicity (5), the above describe all the slices of **KGL**. More precisely, one expects

$$s_n\mathbf{KGL} = \Sigma_{s,t}^n\mathbf{H}\mathbb{Z} .$$

As a consequence, Conjecture 4 and (10) would imply the Atiyah-Hirzebruch type spectral sequence

$$\mathbf{H}^{p-q,q}(X, \mathbb{Z}) \Rightarrow K_{-p-q}(X) . \tag{11}$$

Strong convergence of (11) is shown in [Voe02c]. In the same paper, it was noted that Conjectures 1 and 2 imply Conjecture 4.

5 Appendix

In this appendix, written by the second and third author, we shall discuss in some details the homotopy theoretic underpinnings of the theory presented in the main body of this note.

Some results in motivic homotopy theory depend on a characterization of Nisnevich sheaves in terms of upper distinguished squares. For completeness, in the following section we review the connection between upper distinguished squares and the Nisnevich topology, as described by Morel and Voevodsky in [MV99].

This brings us to the topic of model structures for simplicial presheaves on the smooth Nisnevich site of k. Nowadays there exist several such model structures. At a first encounter, the choice of a model structure might be a quite confusing foundational aspect of the theory. However, the flexibility this choice offers clearly outweighs the drudgery involved in learning about the various models; in fact, this development is a result of various quests to improve the machinery leading to the construction of SH(k). Our exposition follows the paper of Goerss and Jardine [GJ98].

The final section deals with the motivic stable model structure as presented in [Jar00]. In the 1990's topologists discovered model structures with compatible monoidal structures, and such that the associated homotopy categories are all equivalent as monoidal categories to the ordinary stable homotopy category. Motivic stable homotopy theory have in a few years time acquired the same level of technical sophistication as found in ordinary stable homotopy theory. An example of a highly structured model for SH(k) is the model of motivic functors described in [Dun].

In this appendix, using the Zariski spectrum Spec(k) of a field k as the base-scheme is mainly for notational convenience. Although smoothness is essential in the proof of the homotopy purity Theorem 2.26, it is not a foundational requirement for setting up motivic model structures. The theory we shall discuss works well for categories of schemes of (locally) finite type over a finite dimensional Noetherian base-scheme.

5.1 The Nisnevich Topology

In Sect. 2, we constructed several examples of distinguished triangles by means of upper distinguished squares:

$$
\begin{array}{ccc}
W & \longrightarrow & V \\
\downarrow & & \downarrow{\scriptstyle p} \\
U & \xrightarrow{\ i\ } & X
\end{array}
\tag{12}
$$

Recall the conditions we impose on (12): i is an open embedding, p is an etale map, and $p^{-1}(X \smallsetminus U) \to (X \smallsetminus U)$ induces an isomorphism of reduced schemes.

Exercise 5.1 *Show that the real affine line $\mathbb{A}^1_{\mathbb{R}}$, $\mathbb{A}^1_{\mathbb{R}} \smallsetminus \{(x^2+1)\}$, $\mathbb{A}^1_{\mathbb{C}} \smallsetminus \{(x-i)\}$, and $\mathbb{A}^1_{\mathbb{C}} \smallsetminus \{(x+i)(x-i)\}$ define an upper distinguished square. Here, irreducible polynomials are identified with the corresponding closed points on affine lines.*

The Grothendieck topology obtained from the collection of all upper distinguished squares is by definition the smallest topology on Sm/k such that, for each upper distinguished square (12), the sieve obtained from the morphisms i and p is a covering sieve of X, and the empty sieve is a covering sieve of the empty scheme $\emptyset$. Note that for a sheaf F in this topology we have $F(\emptyset) = *$. A sieve of X is a subfunctor of the space represented by X under the Yoneda embedding. A sieve is a covering sieve if and only if it contains a covering arising from an upper distinguished square.

To tie in with the Nisnevich topology, we record the following result due to Morel-Voevodsky [MV99].

Proposition 5.2 *The coverings associated to upper distinguished squares form a basis for the Nisnevich topology on Sm/k.*

In the proof of 5.2, we use the notion of a splitting sequence for coverings: Suppose $X \in \mathrm{Sm}/k$ and the following is a Nisnevich covering

$$\{f_\alpha \colon X_\alpha \longrightarrow X\}_{\alpha \in I} \, . \tag{13}$$

We claim there exists a finite sequence of closed embeddings

$$\emptyset = Z_{\alpha_{n+1}} \longrightarrow Z_{\alpha_n} \longrightarrow \cdots \longrightarrow Z_{\alpha_0} = X \, , \tag{14}$$

and for $0 \leq i \leq n$, $\mathrm{Spec}(k)$-sections s_{α_i} of the natural projections

$$f_{\alpha_i} \times_X (Z_{\alpha_i} \smallsetminus Z_{\alpha_{i+1}}) \colon X_{\alpha_i} \times_X (Z_{\alpha_i} \smallsetminus Z_{\alpha_{i+1}}) \longrightarrow (Z_{\alpha_i} \smallsetminus Z_{\alpha_{i+1}}) \, .$$

To construct the sequence $(Z_{\alpha_i})_{i=0}^{n+1}$ we set $Z_{\alpha_0} := X$. For each generic point x of X, the Nisnevich covering condition requires that there exists an index $\alpha_0 \in I$ and a generic point x_{α_0} of X_{α_0} such that f_{α_0} induces an isomorphism of residue fields $k(x) \to k(x_{\alpha_0})$. The induced morphism of closed integral subschemes corresponding to the generic points is an isomorphism over x, hence an isomorphism over an open neighborhood U_{α_0} of x. It follows that f_{α_0} has a section s_{α_0} over U_{α_0} as shown in the diagram:

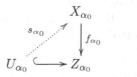

Next, set $Z_{\alpha_1} := Z_{\alpha_0} \smallsetminus U_{\alpha_0}$. With this definition, there exists a Nisnevich covering $\{X_\alpha \times_X Z_{\alpha_1} \to Z_{\alpha_1}\}_{\alpha \in I}$. The next step is to run the same argument for Z_{α_1}. Iterating this procedure, we obtain a strictly decreasing sequence of closed subsets of X with the ascribed property. Since X is Noetherian, the sequence will terminate.

We note that the existence of splitting sequences for coverings implies that the Nisnevich topology on Sm/k is Noetherian in the sense that every covering allows a finite refinement. This follows because there is only a finite number of the pairs $(Z_{\alpha_i} \smallsetminus Z_{\alpha_{i+1}})$ and $\{f_{\alpha_i} : X_\alpha \to X\}_{i=0}^n$ is a Nisnevich covering.

Next, we sketch a proof of 5.2: First, it is clear that every covering obtained from an upper distinguished square is indeed a Nisnevich covering. Conversely, consider the covering sieve R generated by a Nisnevich covering of X (13) and the corresponding splitting sequence (14). Since X is Noetherian, we may assume that I is finite and the sieve R is obtained from the morphism

$$f \colon \coprod_{\alpha \in I} X_\alpha \longrightarrow X \ .$$

The idea is now to construct an upper distinguished square where the scheme subject to the open embedding i has a splitting sequence of length less than that of (14).

Denote by s the $\mathrm{Spec}(k)$-section of f over $X \smallsetminus Z_{\alpha_n}$. Note that we have obtained the upper distinguished square:

$$
\begin{array}{ccc}
W & \longrightarrow & V = (\coprod_{\alpha \in I} X_\alpha) \smallsetminus (f^{-1}(X \smallsetminus Z_{\alpha_n}) \smallsetminus \mathrm{Im}(s)) \\
\downarrow & & \downarrow{\scriptstyle p} \\
U = (X \smallsetminus Z_{\alpha_n}) & \xrightarrow{\quad i \quad} & X
\end{array}
$$

Here, the splitting sequence associated to the Nisnevich covering

$$\coprod_{\alpha \in I} X_\alpha \times_X (X \smallsetminus Z_{\alpha_n}) \longrightarrow (X \smallsetminus Z_{\alpha_n}) \tag{15}$$

has length one less than (14).

The covering sieve R is obtained by composing the Nisnevich coverings

$$\{\coprod_{\alpha \in I} X_\alpha \times_X (X \smallsetminus Z_{\alpha_n}) \longrightarrow (X \smallsetminus Z_{\alpha_n}), V = V\} \ , \tag{16}$$

$$\{U \longrightarrow X, V \longrightarrow X\} \ . \tag{17}$$

By considering the length of the splitting sequence for (15), we may assume that (16) is a covering in the topology generated by upper distinguished squares. The same holds trivially for (17).

Corollary 5.3 *A presheaf on the smooth Nisnevich site of k is a sheaf if and only if it maps every upper distinguished square to a cartesian diagram of sets and the empty scheme to $*$.*

The empty scheme $\emptyset$ represents a simplicial presheaf on Sm/k. Its value on the empty scheme is $*$, not the empty set, which distinguishes it from the initial presheaf. Recall that a Nisnevich neighborhood of $x \in X$ consists of an etale morphism $f \colon V \to X$ together with a point $v \in f^{-1}(x)$ such that the induced map $k(x) \to k(v)$ is an isomorphism. The Nisnevich neighborhoods of $x \in X$ yield a cofiltering system. Let $\mathcal{O}_{X,x}^h$ denote the henselization of the local ring of X at x. Its Zariski spectrum equals the limit of all Nisnevich neighborhoods of $x \in X$.

When F is a sheaf on the smooth Nisnevich site of k, denote by $F(\mathcal{O}_{X,x}^h)$ the filtered colimit of $F(V)$ indexed over all the Nisnevich neighborhoods of x. By restricting to a small skeleton of Sm/k, we obtain a family of conservative points for the Nisnevich topos $\mathrm{Shv}_{\mathrm{Nis}}(\mathrm{Sm}/k)$, namely

$$F \longmapsto F(\mathcal{O}_{X,x}^h) \, .$$

In other terms, a morphism of sheaves $F \to G$ on the smooth Nisnevich site of k is an isomorphism if and only if $F(\mathcal{O}_{X,x}^h) \to G(\mathcal{O}_{X,x}^h)$ is an isomorphism for all $x \in X$.

Denote by $\Delta^{\mathrm{op}}\mathrm{Pre}_{\mathrm{Nis}}(\mathrm{Sm}/k)$ the category of simplicial presheaves on the smooth Nisnevich site of k. Recall from [Dun] the notion of weak equivalence between simplicial sets. A morphism $\mathcal{X} \to \mathcal{Y}$ in $\Delta^{\mathrm{op}}\mathrm{Pre}_{\mathrm{Nis}}(\mathrm{Sm}/k)$ is called a schemewise weak equivalence if for all $X \in \mathrm{Sm}/k$ there is an induced weak equivalence $\mathcal{X}(X) \to \mathcal{Y}(Y)$. In particular, a morphism of discrete simplicial presheaves is a schemewise weak equivalence if and only if it is an isomorphism. There is the much coarser notion of a stalkwise weak equivalence.

Definition 5.4 *A morphism $\mathcal{X} \to \mathcal{Y}$ in $\Delta^{\mathrm{op}}\mathrm{Pre}_{\mathrm{Nis}}(\mathrm{Sm}/k)$ is called a stalkwise weak equivalence if for all $X \in \mathrm{Sm}/k$ and $x \in X$ there is an induced weak equivalence of simplicial sets $\mathcal{X}(\mathcal{O}_{X,x}^h) \to \mathcal{Y}(\mathcal{O}_{X,x}^h)$.*

An important observation is that the simplicial presheaves are evaluated at Hensel local rings; this is particular to the Nisnevich topology. If we instead considered the etale topology, we would have evaluated at strict Hensel local rings. In this way, the stalkwise weak equivalences in $\Delta^{\mathrm{op}}\mathrm{Pre}_{\mathrm{Nis}}(\mathrm{Sm}/k)$ depend on some of the finest properties of the Nisnevich topology.

Exercise 5.5 *Give an example of a stalkwise weak equivalence which is not a weak equivalence on all local rings. (Hint: An example can be obtained by considering the pushout of the upper distinguished square in Exercise 5.1.)*

So far we have encountered two important properties of the Nisnevich topology: First, the collection of all upper distinguished squares generates the Nisnevich topology. This implies a useful characterization of Nisnevich sheaves. Second, the stalkwise weak equivalences are completely determined by Henselizations of Zariski local rings. These two facts are among the chief reasons why developing motivic homotopy theory in the Nisnevich topology turns out to give a whole host of interesting results.

There exist two other characterizations of stalkwise weak equivalences. To review these, we generalize combinatorial and topological homotopy groups for simplicial sets to the setting of Nisnevich sheaves of homotopy groups of simplicial presheaves. Recall that Kan employed the subdivision functor Ex^∞ to define combinatorial homotopy groups of simplicial sets without reference to topological spaces [Dun]. We recall Jardine's generalization to simplicial presheaves [Jar87, §1].

First, we require an extension of Ex^∞ to the simplicial presheaf category $\Delta^{\mathrm{op}}\mathrm{Pre}_{\mathrm{Nis}}(\mathrm{Sm}/k)$. If $\mathcal{X}$ is a simplicial presheaf on the smooth Nisnevich site of k, let $\mathrm{Ex}^m \mathcal{X}$ denote the simplicial presheaf with n-simplices

$$[n] \longmapsto \Delta^{\mathrm{op}}\mathrm{Pre}_{\mathrm{Nis}}(\mathrm{Sm}/k)(\mathrm{sd}^m \Delta[n], \mathcal{X}) \ .$$

In the above expression, sd^m denotes the subdivision functor iterated m times. Its simplicial structure is obtained by precomposition with the simplicial sets maps involving the subdivision sd^m.

Using the natural last vertex maps $\mathrm{sd}\Delta[n] \to \Delta[n]$, for $n \geq 0$, and iterating, we get the diagram

$$\mathcal{X} \longrightarrow \mathrm{Ex}^1 \mathcal{X} \longrightarrow \mathrm{Ex}^2 \mathcal{X} \longrightarrow \cdots \ . \tag{18}$$

Denote by $\mathrm{Ex}^\infty \mathcal{X}$ the colimit of (18) in the presheaf category. There is, by construction, a canonically induced schemewise weak equivalence, and hence stalkwise weak equivalence

$$\mathcal{X} \longrightarrow \mathrm{Ex}^\infty \mathcal{X} \ .$$

A morphism $\mathcal{X} \to \mathcal{Y}$ of simplicial presheaves is a local fibration if for every commutative diagram of simplicial set maps

$$
\begin{array}{ccc}
\Lambda_k[n] & \longrightarrow & \mathcal{X}(X) \\
\downarrow & & \downarrow \\
\Delta[n] & \longrightarrow & \mathcal{Y}(X)
\end{array}
$$

there exists a covering sieve $R \subseteq \mathrm{Sm}/k(-, X)$ such that for every $\phi \colon Y \to X$ in R and in every commutative diagram as below, there exists a lift:

$$
\begin{array}{ccccc}
\Lambda_k[n] & \longrightarrow & \mathcal{X}(X) & \longrightarrow & \mathcal{X}(Y) \\
\downarrow & & & \nearrow & \downarrow \\
\Delta[n] & \longrightarrow & \mathcal{Y}(X) & \longrightarrow & \mathcal{Y}(Y)
\end{array}
$$

Local fibrations are the morphisms with the local right lifting property with respect to the inclusions $\Lambda_k[n] \subseteq \Delta[n]$, $n > 0$, of the boundary $\partial\Delta[n]$ having the k-th face deleted from its list of generators. Simplicial presheaf morphisms having the analogously defined local right lifting property with respect to all inclusions $\partial\Delta[n] \subseteq \Delta[n]$, $n > 0$, are local fibrations. We say that $\mathcal{X}$ is locally fibrant if the morphism $\mathcal{X} \to *$ to the simplicial presheaf represented by $\mathrm{Spec}(k)$ is a local fibration. Schemewise Kan fibrations (i.e. morphisms $\mathcal{X} \to \mathcal{Y}$ which for every member X of Sm/k gives a Kan fibration of simplicial sets $\mathcal{X}(X) \to \mathcal{Y}(X)$) are local fibrations. The simplicial presheaf $\mathrm{Ex}^\infty \mathcal{X}$ is a typical example of a locally fibrant object.

Exercise 5.6 *Show that $\mathcal{X} \to \mathcal{Y}$ is a local fibration if and only if for all $X \in \mathrm{Sm}/k$ and $x \in X$, the map $\mathcal{X}(\mathcal{O}_{X,x}^h) \to \mathcal{Y}(\mathcal{O}_{X,x}^h)$ is a Kan fibration of simplicial sets.*

Conclude that X is locally fibrant if and only if $\mathcal{X}(\mathcal{O}_{X,x}^h)$ is a Kan complex for all $X \in \mathrm{Sm}/k$ and $x \in X$.

When comparing model structures for simplicial presheaves and simplicial sheaves on Sm/k, we shall employ the Nisnevich sheafification functor for presheaves [Lev]. Recall that the functor a_{Nis} is left adjoint to the inclusion

$$\mathrm{Shv}_{\mathrm{Nis}}(\mathrm{Sm}/k) \subseteq \mathrm{Pre}_{\mathrm{Nis}}(\mathrm{Sm}/k).$$

A degreewise application extends it to simplicial presheaves.

Exercise 5.7 *Show that $\mathcal{X} \to a_{\mathrm{Nis}}\mathcal{X}$ is a local fibration by proving that it has the local right lifting property with respect to the inclusions $\partial\Delta[n] \subseteq \Delta[n]$.*

Consider a locally fibrant simplicial presheaf $\mathcal{X}$ on Sm/k and a pair of simplicial set maps

$$\Delta[n] \underset{g}{\overset{f}{\rightrightarrows}} \mathcal{X}(X).$$

Then f is locally homotopic to g if there exists a covering sieve $R \subseteq \mathrm{Sm}/k(-, X)$ such that, for each $\phi: Y \to X$ in R, there is a commutative diagram:

$$
\begin{array}{ccccc}
\Delta[n] & \xrightarrow{\ d^0\ } & \Delta[n] \times \Delta[1] & \xleftarrow{\ d^1\ } & \Delta[n] \\
{\scriptstyle f}\downarrow & & {\scriptstyle h_\phi}\downarrow & & \downarrow{\scriptstyle g} \\
\mathcal{X}(X) & \xrightarrow{\ \phi^*\ } & \mathcal{X}(Y) & \xleftarrow{\ \phi^*\ } & \mathcal{X}(X)
\end{array}
$$

In addition, f and g are locally homotopic relative to $\partial\Delta[n]$ provided each homotopy h_ϕ is constant on $\partial\Delta[n] \subseteq \Delta[n] \times \Delta[1]$ and

$$f|_{\partial\Delta[n]} = g|_{\partial\Delta[n]}.$$

One shows easily that local homotopy relative to $\partial\Delta[n]$ is an equivalence relation for locally fibrant simplicial presheaves [Jar87, Lemma 1.9].

If x is a zero-simplex of $\mathcal{X}(k)$, let $x|X$ be the image of x in $\mathcal{X}(X)_0$ under the canonically induced morphism $\mathcal{X}(\mathrm{Spec}(k)) \to \mathcal{X}(X)$. Consider the following set of all equivalence classes of maps of pairs

$$(\Delta[n], \partial\Delta[n]) \longrightarrow (\mathcal{X}(X), x|X)$$

where the equivalence relation is generated by relative local homotopies. For $n \geq 1$, the associated Nisnevich sheaves $\pi_n^{\mathrm{loc}}(\mathcal{X}, x)$ of combinatorial homotopy groups are formed by letting X vary over the Nisnevich site on Sm/k. When $n = 0$, we take the sheaf associated with local homotopy classes of vertices.

As for simplicial sets, a tedious check reveals that $\pi_n^{\mathrm{loc}}(\mathcal{X}, x)$ is a sheaf of groups for $n \geq 1$, which is abelian for $n \geq 2$.

The Nisnevich site $\mathrm{Sm}/k \downarrow X$ has the terminal object Id_X, with topology induced from the Nisnevich topology on the big site Sm/k. So for a locally fibrant simplicial presheaf $\mathcal{X}$, the zero-simplex $x \in \mathcal{X}|X(\mathrm{Id}_X)_0$ determines a sheaf of homotopy groups $\pi_n^{\mathrm{loc}}(\mathcal{X}|X, x)$.

Definition 5.8 *A morphism $f\colon \mathcal{X} \to \mathcal{Y}$ of simplicial presheaves on the smooth Nisnevich site of k is a combinatorial weak equivalence if for all $n \geq 1$, $X \in \mathrm{Sm}/k$, and zero-simplices $x \in \mathcal{X}(X)_0$ there are induced isomorphisms of Nisnevich sheaves*

$$\pi_0^{\mathrm{loc}}(\mathcal{X}) \longrightarrow \pi_0^{\mathrm{loc}}(\mathcal{Y}) \,,$$

$$\pi_n^{\mathrm{loc}}(\mathrm{Ex}^\infty \mathcal{X}|X, x) \longrightarrow \pi_n^{\mathrm{loc}}(\mathrm{Ex}^\infty \mathcal{Y}|X, f(x)).$$

Exercise 5.9 *Show the following assertions.*

(i) *There is a combinatorial weak equivalence of simplicial presheaves $\mathcal{X} \to \mathcal{Y}$ if and only if for each $X \in \mathrm{Sm}/k$ and $x \in X$ there is a naturally induced weak equivalence of simplicial sets $\mathcal{X}(\mathcal{O}_{X,x}^h) \to \mathcal{Y}(\mathcal{O}_{X,x}^h)$.*

(ii) *If $\mathcal{X}$ is a locally fibrant simplicial presheaf, then $\pi_n^{\mathrm{loc}}(\mathcal{X}) \to \pi_n^{\mathrm{loc}}(\mathrm{Ex}^\infty \mathcal{X})$ is an isomorphism for every $n \geq 0$.*

If $\mathcal{X} \in \Delta^{\mathrm{op}}\mathrm{Pre}_{\mathrm{Nis}}(\mathrm{Sm}/k)$ and $x \in \mathcal{X}(X)_0$ is a zero-simplex, let $\pi_n(\mathcal{X}, x)$ denote the sheaf on $\mathrm{Sm}/k \downarrow X$ associated to the presheaf

$$(U \longrightarrow X) \longmapsto \pi_n(\mathcal{X}(U), x|U) \,.$$

Note that this definition uses homotopy groups obtained by passing to the geometrical realization of the simplicial set $\mathcal{X}(U)$.

The sheaf of path components $\pi_0(\mathcal{X})$ of a simplicial presheaf $\mathcal{X}$ is the Nisnevich sheafification of the coequalizer of the presheaf diagram

$$\mathcal{X}_1 \mathrel{\mathop{\rightrightarrows}^{d_0}_{d_1}} \mathcal{X}_0 \,.$$

The definition of topological weak equivalence is strictly parallel to that of combinatorial weak equivalences:

Definition 5.10 *A morphism $\mathcal{X} \to \mathcal{Y}$ in $\Delta^{\mathrm{op}}\mathrm{Pre}_{\mathrm{Nis}}(\mathrm{Sm}/k)$ is a topological weak equivalence if for all $n \geq 1$, $X \in \mathrm{Sm}/k$, and $x \in \mathcal{X}(X)_0$ there are naturally induced isomorphisms of Nisnevich sheaves*

$$\pi_0(\mathcal{X}) \longrightarrow \pi_0(\mathcal{Y}) \,,$$

$$\pi_n(\mathcal{X}|X, x) \longrightarrow \pi_n(\mathcal{Y}|X, f(x)) \,.$$

For proofs of the following result, see [DI04, 6.7] and [Jar87, 1.18].

Lemma 5.11 *For any simplicial presheaf $\mathcal{X} \in \Delta^{\mathrm{op}}\mathrm{Pre}_{\mathrm{Nis}}(\mathrm{Sm}/k)$, $X \in \mathrm{Sm}/k$, and $x \in \mathcal{X}(X)_0$ there are naturally induced isomorphisms of Nisnevich sheaves*

$$\pi_n(\mathcal{X}|X, x) \longrightarrow \pi_n^{\mathrm{loc}}(\mathcal{X}|X, x) \,.$$

Exercise 5.12 *Let $\mathcal{X} \to \mathcal{Y}$ be a morphism of simplicial presheaves. If $\mathcal{X}$ and $\mathcal{Y}$ are locally fibrant, show that $\mathcal{X} \to \mathcal{Y}$ is a topological weak equivalence if and only if there are naturally induced isomorphisms of Nisnevich sheaves*

$$\pi_0^{\mathrm{loc}}(\mathcal{X}) \longrightarrow \pi_0^{\mathrm{loc}}(\mathcal{Y}) \,,$$

$$\pi_n^{\mathrm{loc}}(\mathcal{X}|X, x) \longrightarrow \pi_n^{\mathrm{loc}}(\mathcal{Y}|X, f(x)).$$

Exercises 5.9 and 5.12 show that the classes of combinatorial, stalkwise, and topological weak equivalences coincide. To emphasize the local structure, we refer to them as local weak equivalences. We also use the notation $\sim$loc.

A simplicial presheaf $\mathcal{X}$ on Sm/k satisfies Nisnevich descent if for every upper distinguished square (12), the following diagram is a homotopy cartesian square of simplicial sets:

$$\begin{array}{ccc} \mathcal{X}(X) & \longrightarrow & \mathcal{X}(V) \\ \downarrow & & \downarrow \\ \mathcal{X}(U) & \longrightarrow & \mathcal{X}(W) \end{array} \qquad (19)$$

The following fundamental result is the Nisnevich descent theorem which was proven by Morel-Voevodsky in [MV99].

Theorem 5.13 *Suppose $\mathcal{X}$ and $\mathcal{Y}$ satisfy Nisnevich descent on Sm/k, and there is a local weak equivalence*

$$\mathcal{X} \xrightarrow{\sim\mathrm{loc}} \mathcal{Y} \,.$$

Then $\mathcal{X}$ and $\mathcal{Y}$ are schemewise weakly equivalent.

Remark 5.14 *We point out that the Nisnevich descent theorem also holds for the big site of finite type S-schemes, where S denotes a Noetherian scheme of finite Krull dimension.*

Theorem 5.13 follows easily from the following Lemma.

Lemma 5.15 *Suppose $\mathcal{X}$ satisfies Nisnevich descent on Sm/k, and there is a local weak equivalence*

$$\mathcal{X} \xrightarrow{\sim \mathrm{loc}} * \, .$$

Then $\mathcal{X}$ and $*$ are schemewise weakly equivalent, so that $\mathcal{X}$ is schemewise contractible.

The Lemma implies the Theorem: Our object is to prove that for every $X \in \mathrm{Sm}/k$, there is a weak equivalence of simplicial sets

$$\mathcal{X}(X) \longrightarrow \mathcal{Y}(X) \, .$$

It suffices to show that the schemewise homotopy fiber over any zero-simplex $x \in \mathcal{Y}(X)_0$ is contractible. Since $\mathcal{X}$ and $\mathcal{Y}$ are locally weakly equivalent and satisfy Nisnevich descent, it follows that the schemewise homotopy fiber is locally weakly equivalent to $*$ and satisfies Nisnevich descent on $\mathrm{Sm}/k \downarrow X$. Lemma 5.15 applies to the Nisnevich site of $\mathrm{Sm}/k \downarrow X$. Thus the homotopy fiber is schemewise contractible. We conclude there is a schemewise weak equivalence

$$\mathcal{X} \xrightarrow{\sim \mathrm{sch}} \mathcal{Y} \, .$$

Later in this text, we shall use an alternate form of the Nisnevich descent theorem in the context of constructing model structures for spectra of spaces. Since the proof of 5.13 and its reformulation makes use of model structures on the presheaf category $\Delta^{\mathrm{op}}\mathrm{Shv}_{\mathrm{Nis}}(\mathrm{Sm}/k)$, we will discuss such model structures in the next section.

5.2 Model Structures for Spaces

This section looks into the construction of a motivic model structure on $\Delta^{\mathrm{op}}\mathrm{Shv}_{\mathrm{Nis}}(\mathrm{Sm}/k)$. Instead of using simplicial sheaves, we shall work in the setting of simplicial presheaves. The motivic model structure for simplicial sheaves follows immediately from the existence of a Quillen equivalent motivic model structure on the corresponding presheaf category $\Delta^{\mathrm{op}}\mathrm{Pre}_{\mathrm{Nis}}(\mathrm{Sm}/k)$.

There are now several model structures underlying the motivic homotopy category. The classes of weak equivalences coincide in all these motivic models. However, the motivic models differ greatly with respect to the choice of cofibrations. Now, the good news is that having a bit a variety in the choice of foundations gives a more in depth understanding of the whole theory. As our cofibrations we choose monomorphisms of simplicial presheaves. This has the neat effect that all objects are cofibrant. On the other hand, this choice makes it difficult to describe the fibrations defined using the right lifting property with respect to trivial cofibrations. In other models, there are more fibrant objects and the fibrations are easier to describe, but then again not every object is cofibrant.

We do not attempt to give a thorough case by case study of each motivic model. The reader can consult the following list of papers on this subject: Blander [Bla01], Dugger [Dug01], Dugger-Hollander-Isaksen [DHI04], Dundas-Röndigs-Østvær [DRØ03], Isaksen [Isa04], Jardine [Jar03], and Voevodsky [Voe00a].

Recall that a cofibration of simplicial sets is simply an inclusion. Jardine [Jar87] proved the theorem that monomorphisms form an adequate class of cofibrations in the simplicial presheaf setting. Adequate means, in particular, that the classes of local weak equivalences and monomorphisms form a model structure on $\Delta^{op}\mathrm{Pre}_{\mathrm{Nis}}(Sm/k)$. This leads to the model structure introduced by Morel-Voevodsky [MV99]. The main innovative idea in their construction of the motivic theory is that the affine line plays the role of the unit interval in topology. In our discussion of the Jardine and Morel-Voevodsky model structures, we follow the approach in the paper of Goerss-Jardine [GJ98].

For the basic notions in homotopical algebra we will use, such as left/right proper and simplicial model categories, see for example [Dun]. First in this section, we deal with Jardine's model structure on $\Delta^{op}\mathrm{Pre}_{\mathrm{Nis}}(Sm/k)$. Global fibrations are, by definition, morphisms having the right lifting property with respect to morphisms which are monomorphisms and local weak equivalences. This forces half of the lifting axiom $\mathcal{M}4$ in the model category structure. We refer to the following model as the local injective model structure.

Theorem 5.16 *The classes of local weak equivalences, monomorphisms and global fibrations define a proper, simplicial and cofibrantly generated model structure for simplicial presheaves on the smooth Nisnevich site of k.*

Fibrant objects in the local injective model structure are called globally fibrant. An essential input in the proof of 5.16 is the following list of properties:

P1 The class of local weak equivalences is closed under retracts.

P2 The class of local weak equivalences satisfies the two out of three axiom.

P3 Every schemewise weak equivalence is a local weak equivalence.

P4 The class of trivial cofibrations is closed under pushouts.

P5 Let γ be a limit ordinal, considered as a partially ordered set. Suppose there is a functor

$$F: \gamma \longrightarrow \Delta^{op}\mathrm{Pre}_{\mathrm{Nis}}(Sm/k) \ ,$$

such that for each morphism $i \leq j$ in γ there is a trivial cofibration

$$F(i) \xrightarrow{\sim \mathrm{loc}} F(j) \ .$$

Then there is a canonically induced trivial cofibration

$$F(i) \xrightarrow{\sim \mathrm{loc}} \operatorname*{colim}_{j \in \gamma} F(j) \ .$$

P6 Suppose there exist trivial cofibrations for $i \in I$

$$F_i \overset{\sim\text{loc}}{\rightarrowtail} G_i \ .$$

Then there is a canonically induced trivial cofibration

$$\coprod_{i \in I} F_i \overset{\sim\text{loc}}{\rightarrowtail} \coprod_{i \in I} G_i \ .$$

P7 There is an infinite cardinal λ which is an upper bound for the cardinality of the set of morphisms of Sm/k such that for every trivial cofibration $\mathcal{X} \overset{\sim\text{loc}}{\rightarrowtail} \mathcal{Y}$ and every λ-bounded subobject $\mathcal{Z}$ of $\mathcal{Y}$ there exists some λ-bounded subobject $\mathcal{W}$ of $\mathcal{Y}$ and a diagram of simplicial presheaves:

$$
\begin{array}{ccc}
\mathcal{W} \cap \mathcal{X} & \hookrightarrow & \mathcal{X} \\
{\scriptstyle\sim\text{loc}}\downarrow & & \downarrow{\scriptstyle\sim\text{loc}} \\
\mathcal{Z} \hookrightarrow \mathcal{W} & \hookrightarrow & \mathcal{Y}
\end{array}
$$

Properties **P1**–**P3** are clear from our discussion of local weak equivalences in Sect. 5.1. For example, the morphism of presheaves of homotopy groups induced by a schemewise weak equivalence is an isomorphism.

To prove **P4**, we consider a pushout diagram in $\Delta^{\text{op}}\text{Pre}_{\text{Nis}}(\text{Sm}/k)$ where i is a cofibration and a local weak equivalence:

$$
\begin{array}{ccc}
\mathcal{X} & \overset{f}{\longrightarrow} & \mathcal{Z} \\
{\scriptstyle i}\downarrow{\scriptstyle\sim\text{loc}} & & \downarrow{\scriptstyle j} \\
\mathcal{Y} & \longrightarrow & \mathcal{Y} \cup_{\mathcal{X}} \mathcal{Z} =: \mathcal{W}
\end{array}
$$

We want to prove that the right vertical morphism is a local weak equivalence. Pushouts along monomorphisms preserves schemewise weak equivalences, so we may assume that all simplicial presheaves are schemewise fibrant; hence, locally fibrant, and moreover that f is a monomorphism. Exercise 5.9 and the characterization of local weak equivalences by combinatorial homotopy groups π^{loc} imply: j is a local weak equivalence if and only if for every $X \in \text{Sm}/k$ and every diagram

$$
\begin{array}{ccc}
\partial\Delta^n & \overset{\alpha}{\longrightarrow} & \mathcal{Z}(X) \\
\downarrow & & \downarrow{\scriptstyle j(X)} \\
\Delta^n & \overset{\beta}{\longrightarrow} & \mathcal{W}(X)
\end{array}
\tag{20}
$$

there exists a covering sieve R of X together with a local homotopy. That is, for every $\phi\colon U \to X$ in R, there is a simplicial homotopy $\Delta^n \times \Delta^1 \to \mathcal{W}(U)$, which is constant on $\partial\Delta^n$, from $\phi^* \circ \beta$ to a map β' with image in $\mathcal{Z}(U)$. Replacing the inclusion $\partial \Delta^n \hookrightarrow \Delta^n$ in diagram (20) by an appropriate subdivision $K \hookrightarrow L$, one can assume that the image under β of every simplex σ of L lies either in $\mathcal{Z}(X)$ or in $\mathcal{Y}(X)$ (or in both, meaning that $\beta(\sigma) \in \mathcal{X}(X) = \mathcal{Y}(X) \cap \mathcal{Z}(X)$). Since L is obtained from K by attaching finitely many simplices (of dimension $0 \le d \le n$), one may construct the required simplicial homotopy by induction on these simplices. In case the simplex has image in $\mathcal{Z}(X)$, use a constant local homotopy. Otherwise, one can construct a local homotopy as desired, because i is a local weak equivalence. Observe that this requires passing to a covering sieve as many times as there are non-degenerate simplices in $L \smallsetminus K$.

The first step in the proof of **P5** is left to the reader as an exercise:

Exercise 5.17 *Note that there is a functor*

$$\mathrm{Ex}^\infty F \colon \gamma \longrightarrow \Delta^{\mathrm{op}} \mathrm{Pre}_{\mathrm{Nis}}(\mathrm{Sm}/k), \quad i \longmapsto \mathrm{Ex}^\infty F(i),$$

together with a natural transformation

$$F \longrightarrow \mathrm{Ex}^\infty F.$$

*By considering the following commutative diagram, show that it suffices to prove **P5** when $F(i)$ is a presheaf of Kan complexes for all $i \in \gamma$:*

$$
\begin{array}{ccc}
F(i) & \longrightarrow & \operatorname*{colim}_{j \in \gamma} F(j) \\
\downarrow & & \downarrow \\
\mathrm{Ex}^\infty F(i) & \longrightarrow & \operatorname*{colim}_{j \in \gamma} \mathrm{Ex}^\infty F(j)
\end{array}
$$

(Hint: Schemewise weak equivalences are local weak equivalences.)

Taking the previous exercise for granted, we may now assume that each $F(i)$ is a presheaf of Kan complexes.

Consider the diagram obtained from the Nisnevich sheafification functor:

$$\begin{array}{ccccc}
F(i) & \longrightarrow & \underset{j\in\gamma}{\mathrm{colim}}\, F(j) & \longrightarrow & a_{\mathrm{Nis}}(\underset{j\in\gamma}{\mathrm{colim}}\, F(j)) \\
\downarrow & & \downarrow & & \downarrow{\scriptstyle\cong} \\
a_{\mathrm{Nis}}F(i) & \longrightarrow & \underset{j\in\gamma}{\mathrm{colim}}\, a_{\mathrm{Nis}}F(j) & \longrightarrow & a_{\mathrm{Nis}}(\underset{j\in\gamma}{\mathrm{colim}}\, a_{\mathrm{Nis}}F(j))
\end{array}$$

Concerning the left lower horizontal morphism, note that $a_{\mathrm{Nis}}F(i) \to a_{\mathrm{Nis}}F(j)$ is a local weak equivalence of locally fibrant simplicial presheaves; hence a schemewise weak equivalence, which implies a schemewise and hence a local weak equivalence between $a_{\mathrm{Nis}}F(i)$ and $\underset{j\in\gamma}{\mathrm{colim}}\, a_{\mathrm{Nis}}F(j)$.

The associated Nisnevich sheaf morphisms are all local weak equivalences, so that starting in the right hand square and using the two out of three property for local weak equivalences, it follows that all the morphisms in the diagram are local weak equivalences. This proves **P5**.

It is time to consider property **P6**. Again, let us start with an exercise.

Exercise 5.18 *Show there is no loss of generality in assuming that F_i and G_i are Kan complexes for all $i \in I$. (Hint: Ex^∞ preserves coproducts.)*

The proof proceeds by noting the local weak equivalence between locally fibrant presheaves $a_{\mathrm{Nis}}F_i \to a_{\mathrm{Nis}}G_i$. In effect, we use

$$\coprod_{i\in I} a_{\mathrm{Nis}}F_i \xrightarrow{\ \sim\mathrm{sch}\ } \coprod_{i\in I} a_{\mathrm{Nis}}G_i \ .$$

Sheafification induces the commutative diagram:

$$\begin{array}{ccccccc}
a_{\mathrm{Nis}}\coprod_{i\in I} F_i & \longleftarrow & \coprod_{i\in I} F_i & \longrightarrow & \coprod_{i\in I} G_i & \longrightarrow & a_{\mathrm{Nis}}\coprod_{i\in I} G_i \\
{\scriptstyle\cong}\downarrow & & \downarrow & & \downarrow & & \downarrow{\scriptstyle\cong} \\
a_{\mathrm{Nis}}\coprod_{i\in I} a_{\mathrm{Nis}}F_i & \longleftarrow & \coprod_{i\in I} a_{\mathrm{Nis}}F_i & \xrightarrow{\ \sim\mathrm{loc}\ } & \coprod_{i\in I} a_{\mathrm{Nis}}G_i & \longrightarrow & a_{\mathrm{Nis}}\coprod_{i\in I} a_{\mathrm{Nis}}G_i
\end{array}$$

By starting with the outer squares, an easy check shows that all morphisms in the diagram are local weak equivalences. The part of property **P6** dealing with cofibrations is clear.

In the formulation of property **P7** or the 'bounded cofibration condition', we implicitly use that Sm/k is skeletally small. The latter means that isomorphism classes of objects in Sm/k form a set. If κ is an infinite cardinal and $X \in \mathrm{Sm}/k$, then the cardinality of X is less than κ, written $\mathbf{card}(X) < \kappa$, if the following hold:

C1 The cardinality of the underlying topological space of X is smaller than the cardinal κ.

C2 For all Zariski open affine patches $\mathrm{Spec}(A)$ of X we have $\mathbf{card}(A) < \kappa$.

Suppose that A is a commutative ring with unit, such that $\mathbf{card}(A) < \kappa$. Then, as a ring, A is isomorphic to the quotient by an ideal of a polynomial ring $\mathbb{Z}[T]$ on a set T of generators such that $\mathbf{card}(T) < \kappa$. This implies the inequality $\mathbf{card}(\mathbb{Z}[T]) < \kappa$. Hence, the cardinality of the collection of ideals of $\mathbb{Z}[T]$ is bounded above by 2^κ. It follows that the collection of isomorphisms classes of all affine schemes $\mathrm{Spec}(A)$ such that $\mathbf{card}(A) < \kappa$ forms a set. To generalize to schemes, use that isomorphism classes of schemes are bounded above by isomorphism classes of diagrams of affine schemes. Fixing an infinite cardinal κ such that $\mathbf{card}(k) < \kappa$, implies, from what we have just observed, that Sm/k is skeletally small; thus, the formulation of **P7** makes sense.

On a related matter, a cofibration in $\Delta^{\mathrm{op}}\mathrm{Pre}_{\mathrm{Nis}}(\mathrm{Sm}/k)$

$$\mathcal{X} \rightarrowtail \mathcal{Y}$$

is λ-bounded if the object $\mathcal{Y}$ is λ-bounded, i.e. for all $X \in \mathrm{Sm}/k$, $n \geq 0$, each set $\mathcal{Y}_n(X)$ has smaller cardinality than λ. For each object $X \in \mathrm{Sm}/k$, there is the X-section functor

$$\mathcal{X} \longmapsto \mathcal{X}(X) \,.$$

It has a left adjoint whose value on the standard simplicial n-simplex $\Delta[n]$ is the λ-bounded simplicial presheaf $h_X\Delta[n]$ defined by

$$Y \longmapsto \coprod\nolimits_{\phi:\, Y \to X} \Delta[n] \,.$$

Using adjointness yields bijections between morphisms of simplicial sets and morphisms of simplicial presheaves,

$$\Delta[n] \longrightarrow \mathcal{Y}(X), h_X\Delta[n] \longrightarrow \mathcal{Y} \,.$$

It follows that any simplicial presheaf on the smooth Nisnevich site of k is a filtered colimit of its λ-bounded subobjects because the generating simplicial presheaves $h_X\Delta[n]$ are all λ-bounded.

Suppose now that $\mathcal{X} \xrightarrow{\sim\mathrm{loc}} \mathcal{Y}$ is given, and choose a λ-bounded subpresheaf $\mathcal{Z} \subseteq \mathcal{Y}$. By applying the functor Ex^∞, we may assume that all simplicial presheaves in sight are locally fibrant. The proof of **P7** proceeds by constructing inductively a sequence of λ-bounded subobjects

$$\mathcal{W}_0 := \mathcal{Z} \subseteq \mathcal{W}_1 \subseteq \mathcal{W}_2 \cdots$$

such that, for each $X \in \mathrm{Sm}/k$, all local lifting problems of the form

$$
\begin{array}{ccc}
\partial\Delta^n & \longrightarrow & \mathcal{W}_i \cap \mathcal{X}(X) \\
\downarrow & & \downarrow \\
\Delta^n & \longrightarrow & \mathcal{W}_i(X)
\end{array}
$$

have solutions over $\mathcal{W}_{i+1}$. Such a local lifting problem amounts to an element e in a relative local homotopy group $\pi_n^{\mathrm{loc}}(\mathcal{W}_i \cap \mathcal{X}, \mathcal{W}_i)$. This element maps to zero in $\pi_n^{\mathrm{loc}}(\mathcal{X}, \mathcal{Y})$. Since local homotopy groups commute with filtered colimits, and since $\mathcal{Y}$ is the filtered colimit of its λ-bounded subobjects by assumption on λ, there exists a λ-bounded subobject $\mathcal{W}_i^e$ such that e maps to zero in the group $\pi_n^{\mathrm{loc}}(\mathcal{W}_i^e \cap \mathcal{X}, \mathcal{W}_i^e)$. The relative local homotopy group $\pi_n^{\mathrm{loc}}(\mathcal{W}_i \cap \mathcal{X}, \mathcal{W}_i)$ is λ-bounded as well, thus $\mathcal{W}_{i+1}$ is the union of all the $\mathcal{W}_i^e$'s. This completes the inductive step.

Set $\mathcal{W} := \cup \mathcal{W}_i$, which is again λ-bounded. It follows, using properties of morphisms having the local right lifting property with respect to the inclusions $\partial \Delta^n \subseteq \Delta^n$, that there is a local weak equivalence

$$\mathcal{W} \cap \mathcal{X} \xrightarrow{\sim \mathrm{loc}} \mathcal{W} .$$

This finishes the sketch proof of **P7**.

In the statement that the local injective model structure is simplicial, we made implicitly use of the fact that the presheaf category $\Delta^{\mathrm{op}}\mathrm{Pre}_{\mathrm{Nis}}(\mathrm{Sm}/k)$ is enriched in the category of simplicial sets. The simplicial structure of a function complex

$$\mathbf{hom}(\mathcal{X}, \mathcal{Y})$$

is determined by

$$\mathbf{hom}(\mathcal{X}, \mathcal{Y})_n := \Delta^{\mathrm{op}}\mathrm{Pre}_{\mathrm{Nis}}(\mathrm{Sm}/k)(\mathcal{X} \times \Delta[n], \mathcal{Y}) .$$

As a simplicial presheaf, the tensor object

$$\mathcal{X} \times \Delta[n]$$

is given by

$$(\mathcal{X} \times \Delta[n])(X) := \mathcal{X}(X) \times \Delta[n] .$$

Pointed function complexes and tensor objects are defined similarly making pointed simplicial presheaves into a category enriched in pointed simplicial sets.

Proof. (Theorem 5.16). In the lectures [Lev] we learned that small limits and small colimits exist for the presheaf category $\mathrm{Pre}_{\mathrm{Nis}}(\mathrm{Sm}/k)$. Hence, the limit axiom $\mathcal{M}1$ holds for $\Delta^{\mathrm{op}}\mathrm{Pre}_{\mathrm{Nis}}(\mathrm{Sm}/k)$ [Dun]. We have already noted that the two out of three axiom $\mathcal{M}2$ holds for the class of local weak equivalences.

The retract axiom $\mathcal{M}3$ holds trivially for both local weak equivalences and cofibrations. Global fibrations are defined by the right lifting property with respect to trivial cofibrations; using this, it follows that global fibrations are closed under retracts.

Consider the lifting axiom $\mathcal{M}4$. In our case, global fibrations are rigged so that the right lifting property part of $\mathcal{M}4$ holds. For the part of $\mathcal{M}4$ which is not true by definition, consider the diagram where i is a cofibration and p is a trivial global fibration:

$$
\begin{array}{ccc}
\mathcal{X}' & \longrightarrow & \mathcal{X} \\
\downarrow & \nearrow & \downarrow {\scriptstyle \sim\text{loc}} \\
\mathcal{Y}' & \longrightarrow & \mathcal{Y}
\end{array}
$$

We want to prove that the indicated filler exists. In the following, let us assume the factorization axiom $\mathcal{M}5$ holds for the canonical morphism

$$\mathcal{Y}' \cup_{\mathcal{X}'} \mathcal{X} \longrightarrow \mathcal{Y}.$$

With this standing assumption, we obtain the commutative diagram:

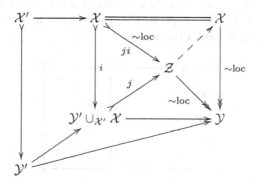

Concerning this diagram we make two remarks:

(i) Note that ji is a cofibration being the composition of the cofibrations.

(ii) Commutativity implies there is a local weak equivalence

$$\mathcal{X} \xrightarrow{\ i\ } \mathcal{Y}' \cup_{\mathcal{X}'} \mathcal{X} \longrightarrow \mathcal{Y}.$$

Hence, there is a local weak equivalence

$$\mathcal{X} \xrightarrow{\ ji\ } \mathcal{Z} \xrightarrow{\ \sim\text{loc}\ } \mathcal{Y}.$$

Thus ji is a trivial cofibration according to $\mathcal{M}2$.

We conclude that the filler with source $\mathcal{Z}$ exists rendering the diagram commutative. This uses the definition of global fibrations in terms of the right lifting property with respect to trivial cofibrations. Note that the above immediately solves our original lifting problem. At this stage of the proof, we have not used the properties **P4–P7**.

The serious part of the proof is to prove the factorization axiom $\mathcal{M}5$. Consider an infinite cardinal λ as in **P7**. We claim that a morphism

is a global fibration if it has the right lifting property with respect to all trivial cofibrations with λ-bounded targets. In other words, for morphisms as above, we claim there exists a filler in every commutative diagram of the form:

$$
\begin{array}{ccc}
\mathcal{X}' & \longrightarrow & \mathcal{X} \\
{\scriptstyle \sim\text{loc}}\Big\downarrow & {\nearrow} & \Big\downarrow \\
\mathcal{Y}' & \longrightarrow & \mathcal{Y}
\end{array}
$$

Here, we may of course assume that the left vertical morphism is not an isomorphism. In effect, there exists a λ-bounded subobject $\mathcal{Z}$ of $\mathcal{Y}'$ which is not a subobject of $\mathcal{X}'$. By property **P7**, there exists a λ-bounded subobject $\mathcal{W}$ of $\mathcal{Y}'$ containing $\mathcal{Z}$, and a diagram:

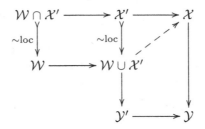

Concerning this diagram we make two remarks:

(i) Property **P4** implies the trivial cofibration

$$\mathcal{X}' \xrightarrow{\;\sim\text{loc}\;} \mathcal{W} \cup \mathcal{X}'.$$

(ii) By the assumption on the right vertical morphism, the partial filler exists.

Consider now the inductively ordered non-empty category of partial lifts where we assume $\mathcal{X}' \neq \mathcal{X}''$:

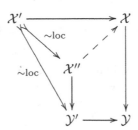

From **P5** and Zorn's lemma, there exists at least one maximal partial lift $\mathcal{X}'' \longrightarrow \mathcal{Y}'$. Maximality implies $\mathcal{X}'' = \mathcal{Y}'$. This solves our lifting problem.

So far, we have made the key observation that morphisms having the right lifting property with respect to all trivial cofibrations with λ-bounded targets are global fibrations. The converse statement holds by definition of global fibrations. We recall from [Dun] that – in more technical terms – this means there exists a set of morphisms, called generating trivial cofibrations, which detects global fibrations. Alas, the argument gives no explicit description of the generators.

We can now set out to construct factorizations of the form:

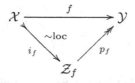

The proof is a transfinite small object argument.

Given a cardinal $\beta > 2^\lambda$ we define inductively a functor

$$F \colon \beta \longrightarrow \Delta^{\mathrm{op}}\mathrm{Pre}_{\mathrm{Nis}}(\mathrm{Sm}/k) \downarrow \mathcal{Y} \,,$$

by setting

(i) $F(0) \colon = f$ and $X(0) = \mathcal{X}$,
(ii) For a limit ordinal ζ,

$$\mathcal{X}(\zeta) \colon = \operatorname*{colim}_{\gamma < \zeta} \mathcal{X}(\gamma) \,.$$

Transitions morphisms are obtained via pushout diagrams

$$
\begin{array}{ccc}
\coprod_D \mathcal{Z}_D & \xrightarrow{\coprod_D i_D} & \coprod_D \mathcal{W}_D \\
\downarrow & & \downarrow \\
\mathcal{X}(\gamma) & \longrightarrow & \mathcal{X}(\gamma + 1)
\end{array}
$$

indexed by the set of all diagrams D where the left vertical morphism is a λ-bounded trivial cofibration:

$$
\begin{array}{ccc}
\mathcal{Z}_D & \xrightarrowtail{\sim\mathrm{loc}} & \mathcal{W}_D \\
\downarrow & & \downarrow \\
\mathcal{X}(\gamma) & \longrightarrow & \mathcal{Y}
\end{array}
$$

We note the following trivial cofibrations:

(i) Property **P6** implies

$$\coprod_D i_D : \coprod_D \mathcal{Z}_D \overset{\sim\text{loc}}{\rightarrowtail} \coprod_D W_D \ .$$

(ii) Part (i) and **P4** imply

$$\mathcal{X}(\gamma) \overset{\sim\text{loc}}{\rightarrowtail} \mathcal{X}(\gamma + 1) \ .$$

Using these constructions, we may now consider the induced factorization:

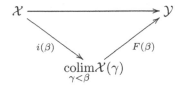

Property **P5** lets us conclude that the first morphism in the factorization is a trivial cofibration.

For the second morphism, one has to solve lifting problems of the form

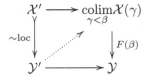

where the left vertical morphism is λ-bounded. To obtain the lifting, note that since $\beta > 2^\lambda$, the upper horizontal morphism factors through some lower stage $\mathcal{X}(\gamma)$ of the colimit.

It remains to prove the second part of the factorization axiom $\mathcal{M}5$. Functorial factorization in the model structure on simplicial sets allows us to factor any morphism in the presheaf category

$$\mathcal{X} \longrightarrow \mathcal{Y}$$

into a cofibration and a schemewise weak equivalence

$$\mathcal{X} \rightarrowtail \mathcal{Z} \overset{\sim\text{sch}}{\longrightarrow} \mathcal{Y} \ .$$

If we factor the schemewise weak equivalence into a trivial cofibration and a fibration, we obtain a commutative diagram:

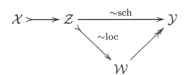

The only comments needed here are:

(i) There is a cofibration obtained by composition of morphisms

$$\mathcal{X} \rightarrowtail \mathcal{W} .$$

(ii) There is a local weak equivalence obtained by **P2** and **P3**

$$\mathcal{W} \xrightarrow{\sim\text{loc}} \mathcal{Y} .$$

Hence items (i) and (ii) imply the desired factorization

$$\mathcal{X} \rightarrowtail \mathcal{W} \xrightarrow{\sim\text{loc}} \mathcal{Y} .$$

An alternate and more honest way of proving the second part of $\mathcal{M}5$ resembles the transfinite small object argument given in the first part. This implies that there exists a set of generating cofibrations.

For the second part of $\mathcal{M}5$, we note a stronger type of factorization result: Given a presheaf morphism

$$\mathcal{X} \longrightarrow \mathcal{Y} ,$$

there exists a factorization

$$\mathcal{X} \rightarrowtail \mathcal{W} \xrightarrow{\sim\text{sch}} \mathcal{Y} .$$

Consider Sm/k in the indiscrete topology, i.e. the only covering sieves are maximal ones [Lev]. One can construct the local injective model structure for the indiscrete topology. This is a simplicial cofibrantly generated model structure on $\Delta^{op}\mathrm{Pre}_{\mathrm{Nis}}(Sm/k)$ where the weak equivalences are schemewise weak equivalences and cofibrations are monomorphisms. We refer to it as the injective model structure.

Applying $\mathcal{M}5$ in the injective model structure to the morphism

$$\mathcal{Z} \xrightarrow{\sim\text{sch}} \mathcal{Y} ,$$

yields the factorization:

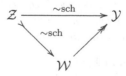

Since $\mathcal{M}2$ holds for schemewise weak equivalences, we immediately obtain the refined form of factorization in the local injective model structure.

Left properness is the assertion that local weak equivalences are preserved under pushouts along cofibrations:

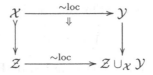

All objects are cofibrant in the local injective model structure. Since in any model category, pushouts of weak equivalences between cofibrant objects along cofibrations are weak equivalences [Hov99, Cube Lemma 5.2.6], left properness follows.

Right properness is the assertion that local weak equivalences are preserved under pullbacks along global fibrations:

$$\begin{array}{ccc} \mathcal{X} \times_{\mathcal{Y}} \mathcal{Z} & \xrightarrow{\sim\text{loc}} & \mathcal{Z} \\ \downarrow & & \downarrow \\ \mathcal{X} & \xrightarrow[\sim\text{loc}]{} & \mathcal{Y} \end{array}$$

However, even a stronger property holds, because local weak equivalences are closed under pullback along local fibrations. The reason is that pullbacks commute with filtered colimits, which implies that it suffices to consider pullback diagrams of the form:

$$\begin{array}{ccc} \mathcal{X} \times_{\mathcal{Y}} \mathcal{Z}(\mathcal{O}^h_{X,x}) & \xrightarrow{\sim} & \mathcal{Z}(\mathcal{O}^n_{X,x}) \\ \downarrow & & \downarrow \\ \mathcal{X}(\mathcal{O}^h_{X,x}) & \xrightarrow{\sim} & \mathcal{Y}(\mathcal{O}^h_{X,x}) \end{array}$$

The result follows because the category of simplicial sets is right proper. □

Making analogous definitions of local weak equivalences, monomorphisms and global fibrations for simplicial sheaves, we infer that there exists a local injective model structure for spaces. The proof consists mostly of repeating arguments we have seen in the simplicial presheaf setting. Details are left to the reader.

Theorem 5.19 *The classes of local weak equivalences, monomorphisms and global fibrations define a proper, simplicial and cofibrantly generated model structure for simplicial sheaves on the smooth Nisnevich site of k.*

Existence of local injective model structures on the categories of pointed simplicial presheaves and pointed simplicial sheaves follow immediately from existence of the respective local injective model structures on $\Delta^{\mathrm{op}}\mathrm{Pre}_{\mathrm{Nis}}(\mathrm{Sm}/k)$ and $\Delta^{\mathrm{op}}\mathrm{Shv}_{\mathrm{Nis}}(\mathrm{Sm}/k)$.

A globally fibrant model for a simplicial presheaf $\mathcal{X}$ consists of a local weak equivalence

$$\mathcal{X} \longrightarrow G\mathcal{X} ,$$

where $G\mathcal{X}$ is globally fibrant. We have seen that globally fibrant models are well-defined up to schemewise weak equivalence. Globally fibrant models exist, and can be chosen functorially, because of the factorization axiom $\mathcal{M}5$. Note, however, that the morphism from $\mathcal{X}$ to $G\mathcal{X}$ is not necessarily a cofibration, so that we will not be tied down to any particular choice of $G\mathcal{X}$. The letter G stands for 'global' or for 'Godement' in the civilized example of the smooth Nisnevich site of k where Godement resolutions yield globally fibrant models. Globally fibrant models in the simplicial sheaf category are defined similarly. Note that a globally fibrant simplicial sheaf is globally fibrant in the simplicial presheaf category.

Exercise 5.20 *Nisnevich descent holds for any globally fibrant simplicial sheaf. (Hint: Use the characterization of sheaves in the Nisnevich topology, see 5.3.)*

We may reformulate the Nisnevich descent Theorem 5.13 in terms of globally fibrant models.

Theorem 5.21 *A simplicial presheaf $\mathcal{X}$ satisfies Nisnevich descent on Sm/k if and only if any globally fibrant model $G\mathcal{X}$ is schemewise weakly equivalent to $\mathcal{X}$.*

Proof. We consider the commutative diagram obtained by sheafifying:

$$
\begin{array}{ccc}
\mathcal{X} & \xrightarrow{\ a_{\mathrm{Nis}}\ } & a_{\mathrm{Nis}}\mathcal{X} \\
\downarrow & & \downarrow \\
G\mathcal{X} & \xrightarrow{\ Ga_{\mathrm{Nis}}\ } & Ga_{\mathrm{Nis}}\mathcal{X}
\end{array}
$$

All morphisms are local weak equivalences. In addition, Ga_{Nis} is a schemewise weak equivalence, since locally weakly equivalent globally fibrant models are schemewise weakly equivalent. Exercise 5.20 implies, since $Ga_{\mathrm{Nis}}\mathcal{X}$ is a sheaf, that $G\mathcal{X}$ satisfies Nisnevich descent.

Theorem 5.13 implies, provided Nisnevich descent holds for $\mathcal{X}$, that there is a schemewise weak equivalence

$$\mathcal{X} \xrightarrow{\ \sim\mathrm{sch}\ } G\mathcal{X} .$$

Conversely, if $\mathcal{X}$ is schemewise weakly equivalent to any of its globally fibrant models; which we have already shown satisfies Nisnevich descent, it follows easily that $\mathcal{X}$ satisfies Nisnevich descent. $\square$

The model structure we shall discuss next is the $\mathbb{A}^1$- or motivic model structure introduced by Morel-Voevodsky [MV99]. Precursors are the local injective model structure and localization techniques developed in algebraic topology. What results is a homotopy theory having deep connections with algebraic geometry.

The motivic model structure arises as the localization theory obtained from the local injective model structure by "formally inverting" any rational point of the affine line

$$* \longrightarrow \mathbb{A}_k^1 \, .$$

Since any two rational points correspond to each other under k-automorphisms of $\mathbb{A}_k^1$, it suffices to consider the zero section $0 \colon \mathrm{Spec}(k) \to \mathbb{A}_k^1$. Getting the motivic theory off the ground involves to a great extend manipulations with function complexes of simplicial presheaves. The main innovative idea is now to replace the local weak equivalences by another class of simplicial presheaf morphisms making the affine line contractible, which we will call motivic weak equivalences, and prove properties **P1-P7** for the motivic weak equivalences. With this input, proceeding as in the construction of the local injective model structure, we get a new cofibrantly generated model structure for simplicial presheaves $\Delta^{\mathrm{op}}\mathrm{Pre}_{\mathrm{Nis}}(\mathrm{Sm}/k)$. This is the motivic model structure.

Definition 5.22 *The classes of motivic weak equivalences and fibrations are defined as follows.*

(i) A simplicial presheaf $\mathcal{Z}$ is motivically fibrant if it is globally fibrant and for every cofibration

$$\mathcal{X} \rightarrowtail \mathcal{Y} \, ,$$

the canonical morphism from $\mathcal{Z}$ to $*$ has the right lifting property with respect to all presheaf inclusions

$$(\mathcal{X} \times \mathbb{A}_k^1) \cup_{\mathcal{X}} \mathcal{Y} \rightarrowtail (\mathcal{Y} \times \mathbb{A}_k^1)$$

induced by the zero section of the affine line.

(ii) A simplicial presheaf morphism

$$\mathcal{X} \longrightarrow \mathcal{Y}$$

is a motivic weak equivalence if for any motivically fibrant simplicial presheaf $\mathcal{Z}$ there is an induced weak equivalence of simplicial sets

$$\mathrm{hom}(\mathcal{Y}, \mathcal{Z}) \longrightarrow \mathrm{hom}(\mathcal{X}, \mathcal{Z}) \, .$$

(iii) A simplicial presheaf morphism is a motivic fibration if it has the right lifting property with respect to morphisms which are simultaneously motivic weak equivalences and monomorphisms.

The lifting property in item (i) is equivalent to having a trivial global fibration

$$\mathbf{Hom}(\mathbb{A}_k^1, \mathcal{Z}) \xrightarrow{\sim^{\mathrm{loc}}} \mathbf{Hom}(*, \mathcal{Z}) \, .$$

Note that the above morphism is always a global fibration.

It follows that a globally fibrant simplicial presheaf $\mathcal{Z}$ is motivically fibrant if and only if all projections

$$\mathbb{A}^1_X \longrightarrow X$$

induce weak equivalences of simplicial sets

$$\mathcal{Z}(X) \overset{\sim}{\longrightarrow} \mathcal{Z}(\mathbb{A}^1_X) .$$

In general, there is no explicit description of motivic fibrations.
The following are examples of motivic weak equivalences:

$$\mathcal{X} \times * \longrightarrow \mathcal{X} \times \mathbb{A}^1_k ,$$

$$(\mathcal{X} \times \mathbb{A}^1_k) \cup_{(\mathcal{X} \times *)} (\mathcal{Y} \times *) \rightarrowtail (\mathcal{Y} \times \mathbb{A}^1_k) .$$

Every local weak equivalence is a motivic weak equivalence for trivial reasons.

Exercise 5.23 *Show that a vector bundle $p \colon X \longrightarrow Y$ in Sm/k is a motivic weak equivalence. Proceed by induction on the number of elements in an open cover of Y which trivializes p.*

Theorem 5.24 *There exists a functor*

$$\mathcal{L} \colon \Delta^{\mathrm{op}}\mathrm{Pre}_{\mathrm{Nis}}(\mathrm{Sm}/k) \longrightarrow \Delta^{\mathrm{op}}\mathrm{Pre}_{\mathrm{Nis}}(\mathrm{Sm}/k) ,$$

and a monomorphism of simplicial presheaves

$$\eta_{\mathcal{X}} \colon \mathcal{X} \longrightarrow \mathcal{L}(\mathcal{X})$$

such that the following holds:

(i) $\mathcal{L}(\mathcal{X})$ is motivically fibrant.
(ii) For every motivically fibrant $\mathcal{Z}$, there is an induced weak equivalence of simplicial sets

$$\mathrm{hom}(\mathcal{L}(\mathcal{X}), \mathcal{Z}) \overset{\sim}{\longrightarrow} \mathrm{hom}(\mathcal{X}, \mathcal{Z}) .$$

In the construction of the functorial motivic fibrant replacement functor $\mathcal{L}$, we shall make use of the fact that there exists a continuous functorial fibrant replacement functor $\mathcal{L}_G$ in the local injective model structure [GJ98]. That $\mathcal{L}_G$ is continuous simply says that the natural maps of hom-sets extend to natural maps of hom-simplicial sets

$$\mathcal{L}_G \colon \mathrm{hom}(\mathcal{X}, \mathcal{Y}) \longrightarrow \mathrm{hom}(\mathcal{L}_G\mathcal{X}, \mathcal{L}_G\mathcal{Y})$$

which are compatible with composition.

Let I be the set of simplicial presheaf morphisms

$$\mathcal{X} \times h_X \Delta[n] \cup_{\mathcal{X} \times \mathcal{Z}} \mathcal{Y} \times \mathcal{Z} \longrightarrow \mathcal{Y} \times h_X \Delta[n] .$$

A typical element in I will be denoted

$$\mathcal{C}_\alpha \longrightarrow \mathcal{D}_\alpha .$$

Choose a cardinal $\beta > 2^\lambda$. Then the first step in the construction of $\mathcal{L}$ is setting

$$\mathcal{L}_0 \mathcal{X} := \mathcal{L}_G \mathcal{X} .$$

At a limit ordinal $\zeta < \beta$, set

$$\mathcal{L}_\zeta \mathcal{X} := \mathcal{L}_G(\operatorname*{colim}_{\gamma < \zeta} \mathcal{L}_\gamma \mathcal{X}) .$$

At successor ordinals, consider the pushout diagram

$$\begin{array}{ccc}
\coprod_{\alpha \in I} \mathcal{C}_\alpha \times \mathbf{hom}(\mathcal{C}_\alpha, \mathcal{L}_\zeta \mathcal{X}) & \longrightarrow & \mathcal{L}_\zeta \mathcal{X} \\
\downarrow & & \downarrow \\
\coprod_{\alpha \in I} \mathcal{D}_\alpha \times \mathbf{hom}(\mathcal{C}_\alpha, \mathcal{L}_\zeta \mathcal{X}) & \longrightarrow & P_I \mathcal{L}_\zeta \mathcal{X}
\end{array}$$

and set

$$\mathcal{L}_{\zeta+1} \mathcal{X} := \mathcal{L}_G(P_I \mathcal{L}_\zeta \mathcal{X}) .$$

These constructions give the natural definition

$$\mathcal{L}\mathcal{X} := \operatorname*{colim}_{\zeta < \beta} \mathcal{L}_\zeta \mathcal{X} .$$

Recall that we choose $\beta > 2^\lambda$ so that any morphism with target $\mathcal{L}\mathcal{X}$ factors through some $\mathcal{L}_\zeta \mathcal{X}$. On account of this observation, we leave it as an exercise to finish the proof of 5.24.

We have the following characterizations of motivic weak equivalences.

Lemma 5.25 *The following assertions are equivalent.*

(i) There is a motivic weak equivalence

$$\mathcal{X} \xrightarrow{\sim \mathrm{mot}} \mathcal{Y} .$$

(ii) For every motivically fibrant simplicial presheaf $\mathcal{Z}$, there is an isomorphism in the local injective homotopy category

$$\mathbf{Ho}_{\Delta^{\mathrm{op}}\mathbf{PreNis}(\mathrm{Sm}/k)_{\sim \mathrm{loc}}}(\mathcal{Y}, \mathcal{Z}) \xrightarrow{\cong} \mathbf{Ho}_{\Delta^{\mathrm{op}}\mathbf{PreNis}(\mathrm{Sm}/k)_{\sim \mathrm{loc}}}(\mathcal{X}, \mathcal{Z}) .$$

(iii) There is a local weak equivalence

$$\mathcal{L}(\mathcal{X}) \xrightarrow{\sim\mathrm{loc}} \mathcal{L}(\mathcal{Y}) \, .$$

Proof. Recall that motivically fibrant objects are in particular globally fibrant, so that (i) implies (ii).

Let $\mathcal{Z}$ be motivically fibrant. By abstract homotopy theory there is an isomorphism in the local injective homotopy category between

$$\mathbf{Ho}_{\Delta^{\mathrm{op}}\mathrm{PreNis}(\mathrm{Sm}/k)_{\sim\mathrm{loc}}}(\mathcal{X}, \mathcal{Z}) \, ,$$

and

$$\mathbf{Ho}_{\Delta^{\mathrm{op}}\mathrm{PreNis}(\mathrm{Sm}/k)_{\sim\mathrm{loc}}}(\mathcal{L}(\mathcal{X}), \mathcal{Z}) \, .$$

When (ii) holds, this implies an isomorphism

$$\mathbf{Ho}_{\Delta^{\mathrm{op}}\mathrm{PreNis}(\mathrm{Sm}/k)_{\sim\mathrm{loc}}}(\mathcal{L}(\mathcal{Y}), \mathcal{Z}) \xrightarrow{\cong} \mathbf{Ho}_{\Delta^{\mathrm{op}}\mathrm{PreNis}(\mathrm{Sm}/k)_{\sim\mathrm{loc}}}(\mathcal{L}(\mathcal{X}), \mathcal{Z}) \, .$$

Theorem 5.24 shows that $\mathcal{L}(\mathcal{X})$ and $\mathcal{L}(\mathcal{Y})$ are motivically fibrant; this implies an isomorphism in the local injective homotopy category

$$\mathcal{L}(\mathcal{X}) \xrightarrow{\cong} \mathcal{L}(\mathcal{Y}) \, .$$

The latter is equivalent to (iii).

When (iii) holds, (i) follows by contemplating the simplicial set diagram:

$$
\begin{array}{ccc}
\mathbf{hom}(\mathcal{L}(\mathcal{Y}), \mathcal{Z}) & \xrightarrow{\sim} & \mathbf{hom}(\mathcal{Y}, \mathcal{Z}) \\
{\scriptstyle\sim}\downarrow & & \downarrow \\
\mathbf{hom}(\mathcal{L}(\mathcal{X}), \mathcal{Z}) & \xrightarrow{\sim} & \mathbf{hom}(\mathcal{X}, \mathcal{Z})
\end{array}
$$

The horizontal morphisms are weak equivalences according to Theorem 5.24. Our assumption implies without much work that the left vertical morphism is a weak equivalence. $\qquad\Box$

We are ready to state the existence of the motivic model structure.

Theorem 5.26 *The classes of motivic weak equivalences, motivic fibrations and monomorphisms define a proper, simplicial and cofibrantly generated model structure for simplicial presheaves on the smooth Nisnevich site of k.*

The proof of 5.26 follows the same script as we have seen for the local injective model structure. Properties **P1-P7** are shown to hold for the class of motivic weak equivalences rather than the class of local weak equivalences. Note that **P1-P3** hold trivially, while **P4-P6** follow from 5.25 using that trivial fibrations of simplicial sets are closed under base change. Finally, the proof of **P7** follows the sketch proof of the same property in the local injective structure, using the motivic fibrant replacement functor [GJ98, 4.7].

Remark 5.27 *In the Morel-Voevodsky paper [MV99] the notions of left and right proper model structures are reversed. However, proper model structure means the usual thing.*

In the following, we discuss the motivic model structure for the category of spaces $\mathrm{Spc}(k)$, i.e. simplicial sheaves on the smooth Nisnevich site of k.

A morphism in $\mathrm{Spc}(k)$ is a motivic weak equivalence if it is a motivic weak equivalence in the simplicial presheaf category. Motivic fibrations are defined similarly. The cofibrations are the monomorphisms.

Concerning sheafified simplicial presheaves and local weak equivalences, there is the following useful result.

Lemma 5.28 *Suppose that $\mathcal{X}$ is a simplicial presheaf and $\mathcal{Y}$ is a simplicial sheaf. Then*

$$\mathcal{X} \longrightarrow \mathcal{Y}$$

is a local weak equivalence if and only if the same holds true for the morphism

$$a_{\mathrm{Nis}}\mathcal{X} \longrightarrow \mathcal{Y} \,.$$

We note the local weak equivalence

$$\mathcal{X} \xrightarrow{\sim\mathrm{loc}} a_{\mathrm{Nis}}\mathcal{X} \,.$$

Since the Nisnevich sheafification functor is idempotent up to isomorphism [Lev], an easy consequence of Lemma 5.28 is that any simplicial presheaf is both local and motivic weakly equivalent to its sheafification.

Exercise 5.29 *Show that a morphism between simplicial sheaves is a motivic fibration if and only if it has the right lifting property with respect to motivic trivial cofibrations of simplicial sheaves.*

Theorem 5.30 *Let $\mathrm{Spc}(k)$ be the category of simplicial Nisnevich sheaves on Sm/k.*

(i) Motivic weak equivalences, motivic fibrations and cofibrations define a proper, simplicial and cofibrantly generated model structure on $\mathrm{Spc}(k)$.

(ii) The Nisnevich sheafification functor induces a Quillen equivalence

$$\Delta^{\mathrm{op}}\mathrm{Pre}_{\mathrm{Nis}}(\mathrm{Sm}/k)_{\sim\mathrm{mot}} \rightleftarrows \Delta^{\mathrm{op}}\mathrm{Shv}_{\mathrm{Nis}}(\mathrm{Sm}/k)_{\sim\mathrm{mot}}.$$

Proof. The limit axiom $\mathcal{M}1$ holds for simplicial sheaves, see e.g. [Lev]. The two out of three axiom $\mathcal{M}2$ and the retract axiom $\mathcal{M}3$ follow immediately since the corresponding statements hold for simplicial presheaves.

Exercise 5.29 shows that the right lifting property part of axiom $\mathcal{M}4$ holds. Given the motivic model structure for simplicial presheaves, the second part of the lifting axiom $\mathcal{M}4$ holds tautologically.

Let $\mathcal{X} \to \mathcal{Y}$ be a morphism in the simplicial sheaf category. We consider the motivic trivial cofibration and motivic fibration factorization part of $\mathcal{M}5$. The motivic model structure shows there is a simplicial presheaf $\mathcal{Z}$ together with morphisms having the required factorization in the simplicial presheaf category. Since $\mathcal{Y}$ is a simplicial sheaf, we may sheafify $\mathcal{Z}$, and employ axiom $\mathcal{M}5$ for the local injective model structure for the simplicial sheaf category. This gives the diagram:

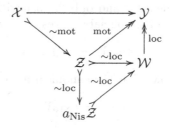

From the above diagram, we deduce:

(i) There is a motivically trivial cofibration of simplicial sheaves

$$\mathcal{X} \xrightarrow{\sim\text{mot}} W .$$

(ii) To obtain the factorization, it suffices to show that the global fibration between W and $\mathcal{Y}$ is a motivic fibration.

As objects of the site $\Delta^{\text{op}}\text{Pre}_{\text{Nis}}(\text{Sm}/k) \downarrow \mathcal{Y}$, note that $\mathcal{Z}$ and W are both cofibrant and globally fibrant. We claim that a local weak equivalence between globally fibrant simplicial presheaves is a schemewise weak equivalence. In fact, a standard trick in simplicial homotopy theory shows the morphism in question is a homotopy equivalence. A global fibration of simplicial sheaves is also a global fibration of simplicial presheaves, and whether a global fibration of simplicial presheaves is also a motivic fibration can be tested schemewise. This implies the statement in (ii).

For the second half of axiom $\mathcal{M}5$, we proceed as above by forming the diagram:

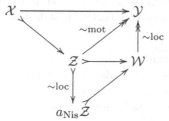

We want to show that the morphism between W and $\mathcal{Y}$ is a motivic fibration. Since the morphism between $\mathcal{Z}$ and $\mathcal{Y}$ is a motivic fibration of simplicial presheaves, we may conclude by noting that the motivic trivial cofibration between $\mathcal{Z}$ and W is a schemewise weak equivalence.

The right adjoint in the adjunction (ii) preserves motivic fibrations and trivial motivic fibrations. Hence we are dealing with a Quillen pair. To show that Nisnevich sheafification is a left Quillen equivalence, let $\mathcal{X}$ be a simplicial presheaf, $\mathcal{Y}$ a motivically fibrant simplicial sheaf, and $\mathcal{Z}$ a motivically fibrant simplicial presheaf. We claim that a morphism in $\Delta^{\mathrm{op}}\mathrm{Pre}_{\mathrm{Nis}}(\mathrm{Sm}/k)_{\sim\mathrm{mot}}$, say

$$\mathcal{X} \longrightarrow \mathcal{Y} \tag{21}$$

is a motivic weak equivalence if and only if Nisnevich sheafification yields a motivic weak equivalence of simplicial sheaves

$$a_{\mathrm{Nis}}\mathcal{X} \longrightarrow \mathcal{Y} . \tag{22}$$

In effect, Lemma 5.28 shows that the canonical map

$$\mathcal{X} \longrightarrow a_{\mathrm{Nis}}\mathcal{X} \tag{23}$$

is a local weak equivalence, hence a motivic weak equivalence. This implies the claim. Moreover, a more refined result holds. Assuming the morphism in (21) is a motivic weak equivalence, we have the diagram of function complexes:

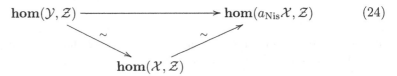

$$\tag{24}$$

Now we use the following facts: The local injective model structure is simplicial, every simplicial presheaf is cofibrant in the local injective model structure, the morphism in (23) is a local weak equivalence, and finally $\mathcal{Z}$ is globally fibrant. Then, by a standard result in homotopical algebra, we get the second weak equivalence indicated in (23). We have just shown for every motivically fibrant simplicial presheaf $\mathcal{Z}$ that there is a weak equivalence

$$\mathbf{hom}(\mathcal{Y}, \mathcal{Z}) \longrightarrow \mathbf{hom}(a_{\mathrm{Nis}}\mathcal{X}, \mathcal{Z}) .$$

In particular, the morphism in (22) is a motivic weak equivalence.

Clearly, the motivic model structur is left proper. For right properness, consider Exercise 5.34. □

Analogously to the work in this section, one shows there exist motivic model structures on the categories of pointed simplicial presheaves and pointed simplicial sheaves on the smooth Nisnevich site of k. The relevant morphisms may be defined via forgetful functors to the unpointed categories.

We will need the following consequences of Nisnevich descent.

Lemma 5.31 *Consider morphisms of motivically fibrant simplicial presheaves*

$$\mathcal{X}_0 \longrightarrow \mathcal{X}_1 \longrightarrow \mathcal{X}_2 \longrightarrow \cdots .$$

(i) *Any globally fibrant model*

$$\underset{n\in\mathbb{N}}{\mathrm{colim}}\ \mathcal{X}_n \longrightarrow G\underset{n\in\mathbb{N}}{\mathrm{colim}}\ \mathcal{X}_n$$

is motivically fibrant.

(ii) *Any motivically fibrant model*

$$\underset{n\in\mathbb{N}}{\mathrm{colim}}\ \mathcal{X}_n \longrightarrow \mathcal{X}$$

is a schemewise weak equivalence.

Proof. Nisnevich descent 5.21 shows that the morphism in (i) is a schemewise weak equivalence. The weak equivalences of simplicial sets

$$\mathcal{X}_n(X) \longrightarrow \mathcal{X}_n(X \times_k \mathbb{A}^1_k)$$

induced from the projection map $X \times_k \mathbb{A}^1_k \to X$ induce weak equivalences in the filtered colimit. This implies (i). Item (ii) follows directly from (i).   □

We will briefly discuss the notion of 'motivic flasque simplicial presheaves'. Such presheaves occupy a central role in the motivic stable theory presented here. Another motivation is that, based in this notion, Isaksen has constructed motivic flasque model structures for simplicial presheaves [Isa04]. First, we shall recall the notion of flasque presheaves. This goes back to the pioneering work of Brown and Gersten on flasque model structures for simplicial sheaves [BG73]. Recently, Lárusson worked in this theme in [Lár04], demonstrating the applicability of abstract homotopy theoretic methods in complex analysis.

A simplicial presheaf $\mathcal{X}$ of Kan complexes is flasque if every finite collection of subschemes X_i of a scheme X induces a Kan fibration

$$\mathbf{hom}(X,\mathcal{X}) \longrightarrow \mathbf{hom}(\cup X_i,\mathcal{X}) .$$

The union is formed in the presheaf category, i.e. $\cup X_i$ is the coequalizer of the diagram of representable presheaves

$$\coprod_{i,j} X_i \times_X X_j \rightrightarrows \coprod_i X_i .$$

There is a canonical monomorphism from $\cup X_i$ to X. In particular, the empty collection of subschemes of X induces the morphism

$$\mathbf{hom}(X,\mathcal{X}) \longrightarrow \mathbf{hom}(\emptyset,\mathcal{X}) .$$

The class of flasque simplicial presheaves is closed under filtered colimits.

Example 5.32 *Globally fibrant simplicial presheaves are flasque.*

Pointed simplicial presheaves $\mathcal{X}$ and $\mathcal{Y}$ have an internal hom

$$\mathbf{Hom}_*(\mathcal{X}, \mathcal{Y}) \,.$$

In X-sections there is the defining equation

$$\mathbf{Hom}_*(\mathcal{X}, \mathcal{Y})(X) = \mathbf{hom}_*(\mathcal{X}|X, \mathcal{Y}|X) \,.$$

Here, the simplicial presheaf $\mathcal{X}|X$ is the restriction of $\mathcal{X}$ to the site $\mathrm{Sm}/k \downarrow X$ along the forgetful functor $\mathrm{Sm}/k \downarrow X \to \mathrm{Sm}/k$. For objects X and Y of Sm/k there is a natural isomorphism

$$\mathbf{Hom}_*(X, \mathcal{Y})(Y) \cong \mathcal{Y}(X \times_k Y) \,.$$

Since the Tate sphere T is the quotient $\mathbb{A}_k^1/(\mathbb{A}_k^1 \smallsetminus \{0\})$, it follows that the internal hom $\mathbf{Hom}_*(T, \mathcal{X})$ sits in the pullback square where the right vertical morphism is induced by the inclusion:

$$
\begin{array}{ccc}
\mathbf{Hom}_*(T, \mathcal{X}) & \longrightarrow & \mathbf{Hom}(\mathbb{A}_k^1, \mathcal{X}) \\
\downarrow & & \downarrow \\
* & \longrightarrow & \mathbf{Hom}((\mathbb{A}_k^1 \smallsetminus \{0\}), \mathcal{X}) \,.
\end{array}
$$

Jardine uses this fact to prove that when $\mathcal{X}$ is flasque, then so is the internal hom $\mathbf{Hom}_*(T, \mathcal{X})$, and moreover, that $\mathbf{Hom}_*(T, -)$ preserves filtered colimits of simplicial presheaves, and schemewise weak equivalences between flasque simplicial presheaves [Jar00, §1.4].

A flasque simplicial presheaf $\mathcal{X}$ is motivically flasque if for all objects $X \in \mathrm{Sm}/k$ the projection

$$X \times_k \mathbb{A}_k^1 \longrightarrow X$$

induces a weak equivalence of simplicial sets

$$\mathcal{X}(X) \longrightarrow \mathcal{X}(X \times_k \mathbb{A}_k^1) \,.$$

If $\mathcal{X}$ is motivically flasque, we have noted that $\mathbf{Hom}_*(T, \mathcal{X})$ is flasque; to see that $\mathbf{Hom}_*(T, \mathcal{X})$ is motivically flasque, it remains to show homotopy invariance. For every $X \in \mathrm{Sm}/k$ we have the fiber sequence

$$\mathbf{Hom}_*(T, \mathcal{X})(X) \longrightarrow \mathcal{X}(X \times_k \mathbb{A}_k^1) \longrightarrow \mathcal{X}(X \times_k (\mathbb{A}_k^1 \smallsetminus \{0\})) \,.$$

Comparing with the corresponding fiber sequence for $X \times_k \mathbb{A}_k^1$, it follows that $\mathbf{Hom}_*(T, \mathcal{X})$ is homotopy invariant.

The next lemma summarizes some properties of the internal hom functor $\mathbf{Hom}_*(T, -)$.

Lemma 5.33 *The Tate sphere satisfies the following properties.*

(i) For sequential diagrams of pointed simplicial presheaves, we have

$$\mathbf{Hom}_*(T, \operatorname*{colim}_{n \in \mathbb{N}} \mathcal{X}_n) \cong \operatorname*{colim}_{n \in \mathbb{N}} \mathbf{Hom}_*(T, \mathcal{X}_n) \, .$$

(ii) If $\mathcal{X}$ is motivically flasque, then so is the internal hom $\mathbf{Hom}_*(T, \mathcal{X})$.

(iii) $\mathbf{Hom}_*(T, -)$ preserves schemewise equivalences between motivic flasque simplicial presheaves.

That the Tate sphere is 'compact' refers to the combination of all the properties listed in 5.33.

Modern formulations of homotopical algebra allow for different approaches to the local injective and motivic model structures. One of these approaches is via Bousfield localization. In the context of cellular model categories, the authorative reference on this subject is Hirschhorn's book [Hir03]. The notion of combinatorial model categories, as introduced by Jeff Smith [Smi], provide acceptable inputs for Bousfield localization. A model category is combinatorial if the model structure is cofibrantly generated, and the underlying category is locally presentable. This makes it quite plausible that all the model structures on simplicial presheaves on the smooth Nisnevich site of k that we discussed are indeed combinatorial. Starting with the schemewise model structure, with schemewise weak equivalences and schemewise cofibrations, the local and the motivic model structure can be constructed using Bousfield localizations. In general, right properness is not preserved under Bousfield localization of model structures. However, the motivic model structure is right proper.

Exercise 5.34 *Compare the proofs of right properness of the motivic model structure in [Jar00, Appendix A] and [MV99, Theorem 2.7].*

This finishes our synopsis of basic motivic unstable homotopy theory.

5.3 Model Structures for Spectra of Spaces

This section deals with the nuts and bolts of the model structures underlying the motivic stable homotopy theory introduced by Voevodsky [Voe98].

The original reference for the material presented in this section is [Jar00]. We will not attempt to cover the motivic symmetric spectra part of Jardine's paper. The main point of working with the category of motivic symmetric spectra is that it furnishes a model category for the motivic stable homotopy theory with an internal symmetric monoidal smash product. These issues, and some other deep homotopical structures, are discussed from an enriched functor point of view in [Dun]. Using a Quillen equivalent model structure for the motivic unstable homotopy category, Hovey [Hov01] constructed a model structure similar to the one we will discuss here. A major difference between the approaches in [Hov01] and in [Jar00] is that Hovey does not use the Nisnevich descent theorem in the construction of the model structure. At any rate, using the internal smash product for symmetric spectra and comparing with ordinary spectra, it follows without much fuss that $SH(k)$ has the structure of a closed symmetric monoidal and triangulated category. The homotopy categories $SH_s(k)$ and $SH_s^{\mathbb{A}^1}(k)$ acquire the exact same type of structure.

First, we discuss the level model structures, and second the stable model structure. There are two level model structures. These structures share the same class of weak equivalences, but their classes of cofibrations and fibrations do not coincide. This is reminiscent of the situation with different models for the motivic unstable homotopy category. The interplay between the level models are important for the construction of the more interesting stable model structure, whose associated homotopy category is the motivic stable homotopy category.

The motivic spectra we consider are suspended with respect to the Tate sphere T, i.e. sequences of pointed simplicial presheaves $E = \{\mathcal{E}_n\}_{n \geq 0}$ on the smooth Nisnevich site of k together with structure maps

$$\sigma \colon T \wedge \mathcal{E}_n \longrightarrow \mathcal{E}_{n+1} \; .$$

The usual compatibility conditions are required for morphisms of motivic spectra. Note that, in the smash product, the Tate sphere is placed on the left hand side.

An optimistic, but homotopy theoretic correct definition of the smash product of two motivic spectra is given by

$$(E \wedge E')_n \; := \; \begin{cases} \mathcal{E}_i \wedge \mathcal{E}_i' & n = 2i, \\ T \wedge (\mathcal{E}_i \wedge \mathcal{E}_i') & n = 2i + 1. \end{cases}$$

In even degrees, the structure map is the identity, while in degrees $n = 2i + 1$ one makes the choice

$$T \wedge (T \wedge (\mathcal{E}_i \wedge \mathcal{E}_i')) \xrightarrow{\cong} (T \wedge \mathcal{E}_i) \wedge (T \wedge \mathcal{E}_i') \xrightarrow{\sigma \wedge \sigma'} \mathcal{E}_{i+1} \wedge \mathcal{E}_{i+1}' \; .$$

Then the following diagram commutes, where, up to sign, the left vertical twist isomorphism is homotopic to the identity:

$$T \wedge (T \wedge (\mathcal{E}_i \wedge \mathcal{E}'_i)) \xrightarrow[\text{(23)}]{\cong} (T \wedge \mathcal{E}_i) \wedge (T \wedge \mathcal{E}'_i) \longrightarrow (T \wedge \mathcal{E}_i) \wedge \mathcal{E}'_{i+1}$$

$$\cong \Big\downarrow \text{(12)}$$

$$T \wedge (T \wedge (\mathcal{E}_i \wedge \mathcal{E}'_i))$$

$$\|$$

$$T \wedge ((T \wedge \mathcal{E}_i) \wedge \mathcal{E}'_i)$$

$$\downarrow$$

$$T \wedge (\mathcal{E}_{i+1} \wedge \mathcal{E}'_i) \xrightarrow{\cong} \mathcal{E}_{i+1} \wedge (T \wedge \mathcal{E}'_i) \longrightarrow \mathcal{E}_{i+1} \wedge \mathcal{E}'_{i+1}$$

It follows that the suggested smash product of spectra is neither associative nor commutative before passing to the homotopy category. Hence, the smash products and actions are only given up to homotopy. See also Remark 2.14.

There are some set theoretic problems involved in inverting a class of morphisms in a category. Once the model structure has been constructed, we may define the motivic stable homotopy category. Quillen's theory of model structures, or homotopical algebra, provides the foundation for any treatment of motivic stable homotopy theory.

In the motivic levelwise model structures, the weak equivalences are levelwise motivic weak equivalences. We may choose levelwise cofibrations or levelwise fibrations. Both choices induce model structures on motivic spectra.

To construct the motivic stable model structure, we employ the T-loops functor. It is right adjoint to smashing with the Tate sphere functor. This leads to the process of T-stabilization, and a proof of the model axioms for motivic spectra which avoids reference to Nisnevich sheaves of homotopy groups. The cofibrations are defined levelwise. We end the discussion by relating motivic stable weak equivalences to Nisnevich sheaves of bigraded stable homotopy groups of (s, t)-spectra, as in Sect. 2.3.

Although the construction of the motivic stable model structure is more involved, we note that formal techniques originating in the study of spectra of simplicial sets can be hoisted to motivic spectra. We will make use of the approach set forth by Bousfield-Friedlander [BF78], and of injective motivic spectra as a notion for fibrant objects in the motivic level model structure; the latter uses ideas introduced in the Hovey-Shipley-Smith paper on symmetric spectra of simplicial sets [HSS00].

From now on, all simplicial presheaves are pointed.

Definition 5.35 *A morphism of motivic spectra*

$$E \longrightarrow E'$$

is a levelwise equivalence if for every non-negative integer $n \geq 0$, there is a motivic weak equivalence of simplicial presheaves

$$\mathcal{E}_n \xrightarrow{\sim \mathrm{mot}} \mathcal{E}'_n \ .$$

Levelwise cofibrations and levelwise fibrations are defined likewise.

A cofibration is a morphism having the left lifting property with respect to all levelwise equivalences which are levelwise fibrations.

An injective fibration is a morphism having the right lifting property with respect to all levelwise equivalences which are levelwise cofibrations.

Lemma 5.36 *Let $n \geq 1$ and consider a morphism of motivic spectra*

$$i \colon E \longrightarrow E' \ ,$$

having the additional properties that there are canonically induced cofibrations of simplicial presheaves on the smooth Nisnevich site of k

$$\mathcal{E}_0 \rightarrowtail \mathcal{E}'_0 \ ,$$

$$\mathcal{E}_n \cup_{T \wedge \mathcal{E}_{n-1}} T \wedge \mathcal{E}'_{n-1} \rightarrowtail T \wedge \mathcal{E}'_n.$$

Then i is a cofibration of motivic spectra.

Proof. Consider the lifting problems where the right hand vertical morphism in the diagram of motivic spectra is a levelwise equivalence and levelwise fibration:

$$
\begin{array}{ccc}
E \longrightarrow F & \quad & \mathcal{E}_n \longrightarrow \mathcal{F}_n \\
\downarrow^{i} \ \ \nearrow^{s} \ \ \downarrow & \quad & \downarrow^{i_n} \ \ \nearrow^{s_n} \ \ \downarrow \\
E' \longrightarrow F' & \quad & \mathcal{E}'_n \longrightarrow \mathcal{F}'_n
\end{array}
$$

We construct fillers s_n and s by using an induction argument.

If $n = 0$, then since the right hand vertical morphism is a motivically trivial fibration and cofibrations in the motivic model structure are monomorphisms, the filler s_0 exists according to axiom $\mathcal{M}4$ for the motivic model structure.

Suppose that the n-th filler s_n has been constructed. Then, since we are dealing with morphisms of spectra, there is the commutative diagram:

$$
\begin{array}{ccc}
T \wedge \mathcal{E}_n \longrightarrow T \wedge \mathcal{E}'_n \xrightarrow{\Sigma_T s_n} T \wedge \mathcal{F}_n \\
\downarrow \qquad\qquad\qquad\qquad\qquad\qquad \downarrow \\
\mathcal{E}_{n+1} \longrightarrow\qquad\qquad\qquad\qquad \mathcal{F}_{n+1}
\end{array}
$$

This allows us to consider the commutative diagram:

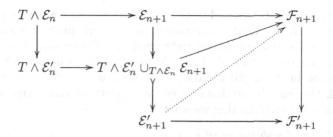

Now, the lower central vertical morphism is a cofibration by our assumptions, which implies, using the argument for $n = 0$, that the indicated filler exists. By commutativity of the diagram this morphism is also the $n + 1$-th filler.

We leave it to the reader to verify the fact that the fillers assemble into a morphism of motivic spectra.   □

Exercise 5.37 *With the same notations and assumptions as in the previous Lemma, show that if the cofibrations of simplicial presheaves are motivic weak equivalences, then i is a level equivalence and cofibration.*

We have collected the crux ingredients needed in the proof of:

Proposition 5.38 *The category of motivic spectra together with the classes of level equivalences, cofibrations, and level fibrations has the structure of a proper simplicial model category.*

The simplicial model structure arises from the smash products $E \wedge K$, where K is a pointed simplicial set, and the function complexes $\mathbf{hom}_*(E, E')$ with n-simplices all morphisms $E \wedge \Delta[n]_-+ \rightarrow E'$. In this definition, we consider the standard n-simplicial set with an added disjoint base-point as a constant pointed space.

Suppose we want to factor a morphism of motivic spectra

$$E \longrightarrow E'$$

into a cofibration and a level equivalence, followed by a level fibration.

In level zero, this follows from the motivic model structure.

Assume there exist such factorizations up to level n, and consider the commutative diagram:

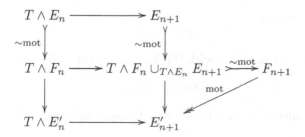

The left vertical cofibration and motivic weak equivalence are both part of the induction hypothesis. Trivial cofibrations are closed under pushouts in any model category, so that we get the right vertical cofibration and motivic weak equivalence. There is a canonical morphism from the pushout to E'_{n+1} which we may factor in the motivic model structure, as depicted in the diagram. Using 5.36, this clearly produces a motivic spectrum consisting of the F_i's together with the factorization we wanted.

Exercise 5.39 *Finish the proof of 5.38.*

There is the following analogous result involving injective fibrations. We will not dwell into the details of the proof, which uses a transfinite small object argument, since the techniques are reminiscent of what we have seen for simplicial presheaves. The method of proof is to show for motivic spectra properties analogous of **P1-P7**.

Proposition 5.40 *The category of T-spectra together with the classes of level equivalences, level cofibrations, and injective fibrations is a proper simplicial model category.*

If $\mathcal{X}$ is a simplicial presheaf on Sm/k, its T-loops functor is defined by setting

$$\Omega_T \mathcal{X} := \mathbf{Hom}_*(T, \mathcal{X}) .$$

Taking T-loops is right adjoint to smashing with T.

The T-loops $\Omega_T E$ of a motivic spectrum E is defined by setting

$$(\Omega_T E)_n := \Omega_T \mathcal{E}_n$$
$$= \mathbf{Hom}_*(T, \mathcal{E}_n) .$$

The structure maps

$$\sigma_T : T \wedge (\Omega_T E)_n \longrightarrow (\Omega_T E)_{n+1}$$

are defined, and this is a possible source for confusion, by taking the adjoint of the composite morphism

$$T \wedge \Omega_T \mathcal{E}_n \wedge T \longrightarrow T \wedge \mathcal{E}_n \longrightarrow \mathcal{E}_{n+1} .$$

The T-loops functor

$$E \longmapsto \Omega_T E$$

is right adjoint to smashing with the Tate sphere on the right

$$E \longmapsto E \wedge T .$$

This gives an alternate way of describing the structure maps of $\Omega_T E$ by taking adjoints

$$\sigma_T^* \colon \mathcal{E}_n \longrightarrow \mathbf{Hom}_*(T, \mathcal{E}_{n+1}) \,.$$

There is another functor Ω_T^ℓ which Jardine calls the 'fake T-loop functor' [Jar00]. By definition, there is the T-spectrum

$$(\Omega_T E)_n = (\Omega_T^\ell E)_n \,,$$

with structure maps adjoint to the morphisms obtained by applying Ω_T to σ_T^*. That is,

$$\Omega_T(\sigma_T^*) \colon \Omega_T(\mathcal{E}_n) \longrightarrow \Omega_T(\mathbf{Hom}_*(T, \mathcal{E}_{n+1})) \,.$$

The reason for the letter ℓ is that the fake T-loops functor

$$E \longmapsto \Omega_T^\ell E$$

is right adjoint to smashing with the Tate sphere on the left

$$E \longmapsto T \wedge E \,,$$

where $\sigma_n^{T \wedge E} = T \wedge \sigma_n^E$.

Suppose that m is an integer and E is a motivic spectrum. Then a shifted motivic spectrum $E[m]$ is obtained, in the range where it makes sense, by setting

$$E[m]_n \; : \; = \begin{cases} \mathcal{E}_{m+n} & m+n \geq 0, \\ * & m+n < 0. \end{cases}$$

The structure maps are reindexed accordingly. Note that $E[m]$ gives iterated suspensions when m is positive, and iterated loops when m is negative.

Exercise 5.41 *Let E be a motivic spectrum. Is it true that the morphisms*

$$(E \wedge T)_n = \mathcal{E}_n \wedge T \overset{\cong}{\longrightarrow} T \wedge \mathcal{E}_n \to \mathcal{E}_{n+1} = E[1]_n$$

form a morphism of motivic spectra between $E \wedge T$ and $E[1]$?

The case $m = 1$ is particularly important because the morphisms σ_T^* determine a morphism of motivic spectra

$$\sigma_T^* \colon E \longrightarrow \Omega_T^\ell E[1] \,.$$

By iterating the above, as many times as there are natural numbers, we get the sequence

$$E \xrightarrow{\;\sigma_T^*\;} \Omega_T^\ell E[1] \xrightarrow{\;\Omega_T^\ell \sigma_T^*[1]\;} (\Omega_T^\ell)^2 E[2] \xrightarrow{\;(\Omega_T^\ell)^2 \sigma_T^*[2]\;} \cdots \,. \qquad (25)$$

Let $Q_T E$ denote the colimit of the diagram (25), and consider the canonically induced morphism

$$\eta_E \colon E \longrightarrow Q_T E$$

The functor Q_T is called the stabilization functor for the Tate sphere.

We will have occasions to consider the level fibrant model of E obtained from the motivic levelwise model structure

$$j_E \colon E \longrightarrow JE\,,$$

and the composite morphism

$$\tilde\eta_E \colon E \xrightarrow{\;j_E\;} JE \xrightarrow{\;\eta_{JE}\;} Q_T JE\,.$$

We are ready to define stable equivalences and stable fibrations.

Definition 5.42 *Let*

$$\phi\colon E \longrightarrow E'$$

be a morphism of motivic spectra. Then

(i) ϕ *is a stable equivalence if it induces a level equivalence*

$$Q_T J(\phi)\colon Q_T JE \longrightarrow Q_T JE'\,.$$

(ii) ϕ *is a stable fibration if it has the right lifting property with respect to all morphisms which are cofibrations and stable equivalences.*

A first observation is

Lemma 5.43 *Level equivalences are stable equivalences.*

Proof. Taking the level fibrant model of a level equivalence yields a level equivalence between level fibrant motivic spectra. Now, in each level there is a motivic weak equivalence of motivically fibrant objects, so that a standard argument for simplicial model categories shows that we are dealing with a schemewise weak equivalence of motivically flasque objects. We may conclude since the Tate sphere is compact according to 5.33. □

Lemma 5.43 shows that every stable fibration is a levelwise fibration.

The main theorem in this section is:

Theorem 5.44 *The category of motivic spectra together with the classes of stable equivalences, cofibrations, and stable fibrations forms a proper simplicial model category.*

In the proof, we make use of the following Lemma.

Lemma 5.45 *Let E and E' be motivic spectra.*

(i) A levelwise fibration

$$\phi\colon E \longrightarrow E'$$

is a stable fibration if there is a level homotopy cartesian diagram:

$$
\begin{array}{ccc}
E & \longrightarrow & Q_T J E \\
\phi \downarrow & & \downarrow Q_T J(\phi) \\
E' & \longrightarrow & Q_T J E'
\end{array}
$$

(ii) Stable equivalences are closed under pullbacks along level fibrations.

(iii) If E is stably fibrant, then E is level fibrant, and there are schemewise weak equivalences

$$\sigma_T^*\colon \mathcal{E}_n \longrightarrow \mathbf{Hom}_*(T, \mathcal{E}_{n+1}).$$

In particular, from (i) and (iii), we note that E is stably fibrant if and only if E is level fibrant and for all $n \geq 0$, there are schemewise equivalences

$$\sigma_T^*\colon \mathcal{E}_n \longrightarrow \mathbf{Hom}_*(T, \mathcal{E}_{n+1}).$$

Proof. Given 5.43, item (i) follows provided there are levelwise equivalences

$$Q_T J(\tilde{\eta}_E) = Q_T J(\eta_{JE}) \circ Q_T J(j_E)\colon Q_T J E \longrightarrow Q_T J^2 E \longrightarrow (Q_T J)^2 E$$

$$\tilde{\eta}_{Q_T JE} = \eta_{J Q_T JE} \circ j_{Q_T JE}\colon Q_T J E \longrightarrow J Q_T J E \longrightarrow (Q_T J)^2 E.$$

Let us consider the first level equivalence. The morphism $Q_T J(j_E)$ is a level equivalence by construction. To show that $Q_T J(\eta_{JE})$ is a level equivalence, we consider the commutative diagram:

$$
\begin{array}{ccc}
Q_T J E & \xrightarrow{Q_T(\eta_{JE})} & Q_T Q_T J E \\
Q_T(j_E) \downarrow & & \downarrow Q_T(j_{Q_T JE}) \\
Q_T J^2 E & \xrightarrow{Q_T J(\eta_{JE})} & (Q_T J)^2 E
\end{array}
$$

A cofinality argument shows $Q_T(\eta_{JE})$ is an isomorphism since $\mathbf{Hom}_*(T, -)$ commutes with sequential colimits of pointed simplicial presheaves; cp. 5.33. Partially by definition, the morphism j_{JE} is a level equivalence which in each level consists of motivically flasque simplicial presheaves. Lemma 5.33, see (i) and (ii), shows that stabilizing with respect to the Tate sphere T preserves this property, so that $Q_T(j_{JE})$ is a level equivalence.

Now the crux of the proof is that Nisnevich descent and compactness of the Tate sphere imply that the right vertical morphism is a level equivalence, see 5.31(ii), applied to $j_{Q_T JE}$, and 5.33(i), (ii). The above implies that the composition $Q_T J(\tilde{\eta}_E)$ is a level equivalence.

Next, we consider the second level equivalence. We have already noted the left vertical schemewise weak equivalence and upper horizontal isomorphism in the commutative diagram:

$$
\begin{array}{ccc}
(Q_T JE)_n & \xrightarrow{\sigma_T^*} & \mathbf{Hom}_*(T, (Q_T JE)_{n+1}) \\
{\scriptstyle j_{Q_T JE}}\big\downarrow & & \big\downarrow{\scriptstyle \mathbf{Hom}_*(T, j_{Q_T JE})} \\
(Q_T JE)_{n+1} & \xrightarrow{\sigma_T^*} & \mathbf{Hom}_*(T, (JQ_T JE)_{n+1})
\end{array}
$$

Lemma 5.33(i), and (ii) applied to $j_{Q_T JE}$ implies the level equivalence

$$
\eta_{JQ_T JE} \colon JQ_T JX \longrightarrow Q_T JQ_T JX \ .
$$

This implies the level equivalence $\tilde{\eta}_{Q_T JE}$.

Item (ii) follows from properness of the motivic levelwise model structure with level fibrations, together with a straight-forward argument.

In the proof of (iii), we employ the motivic levelwise model structure with injective fibrations. There is a natural level cofibration and level equivalence of motivic spectra

$$
i_E \colon E \longrightarrow IE,
$$

where IE is injective. Generally, a level equivalence with an injective target is called an injective model for the source.

Exercise 5.46 *Show that a morphism of motivic spectra*

$$
\phi \colon E \longrightarrow E'
$$

is a stable equivalence if and only if it induces a level equivalence

$$
IQ_T J(\phi) \colon IQ_T JE \longrightarrow IQ_T JE' \ .
$$

Show that $IQ_T JE$ is stably fibrant.

We have shown that the composite morphism

$$
E \xrightarrow{\ j_E\ } JE \xrightarrow{\ \eta_{JE}\ } Q_T JE \xrightarrow{\ i_{Q_T JE}\ } IQ_T JE
$$

is a stable equivalence. We may factor the latter morphism into a cofibration, followed by a level equivalence and level fibration:

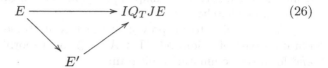

$$E \longrightarrow IQ_T JE \qquad (26)$$

Note that the morphism

$$E' \longrightarrow IQ_T JE$$

is a stable fibration because it has the right lifting property with respect to all cofibrations. Hence E' is stably fibrant, and there are schemewise weak equivalences in all levels

$$\sigma_T^*\colon \mathcal{E}_n' \longrightarrow \mathbf{Hom}_*(T, \mathcal{E}_{n+1}').$$

Moreover, the cofibration in diagram (26) is also a stable equivalence according to 5.43 and the two out of three property of stable equivalences, so that E is a retract of E'. The result follows.   □

Corollary 5.47 *The following hold.*

(i) *A morphism of motivic spectra is a stable fibration and stable equivalence if and only if it is a level fibration and a level equivalence.*

(ii) *Every level fibration between two stably fibrant motivic spectra is a stable fibration.*

Following the script for ordinary spectra, our aim is now to finish the proof of 5.44. There is really only axiom $\mathcal{M}5$ which requires a comment.

Proof. First, we note that the category of motivic spectra

$$\mathrm{Spt}(\Delta^{\mathrm{op}}\mathrm{Pre}_{\mathrm{Nis}}(\mathrm{Sm}/k), T)$$

is bicomplete: If F is a functor from a small category I to motivic spectra, one puts

$$(\operatorname*{colim}_{i \in I} F(i))_n := \operatorname*{colim}_{i \in I} (F(i))_n,$$

$$(\operatorname*{lim}_{i \in I} F(i))_n := \operatorname*{lim}_{i \in I} (F(i))_n .$$

When forming colimits, the structure maps are given by

$$\operatorname*{colim}_{i \in I} \sigma : T \wedge (\operatorname*{colim}_{i \in I} (F(i))_n) \cong \operatorname*{colim}_{i \in I} (T \wedge F(i)_n) \longrightarrow \operatorname*{colim}_{i \in I} (F(i)_n) .$$

The isomorphism we use above arises from the canonical morphism from the colimit of the suspension with T functor to the same functor applied to the colimit; since the suspension is a left adjoint – which we have inverted – this is an isomorphism. When forming limits, the structure maps are defined similarly using the adjoint structure maps σ_T^*. Axiom $\mathcal{M}1$ for $\mathrm{Spt}(\Delta^{\mathrm{op}}\mathrm{Pre}_{\mathrm{Nis}}(\mathrm{Sm}/k), T)$ follows immediately.

What remains to be proven is the trivial stable cofibration and stable fibration part of axiom $\mathcal{M}5$. Let $\mathcal{X} \to \mathcal{Y}$ be a morphism of motivic spectra and form the commutative diagram:

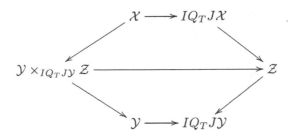

The right hand side makes use of the cofibration–level equivalence and level fibration factorization axiom. Then, $\mathcal{Z}$ is level fibrant and in each level, the cofibration–level equivalence is a schemewise equivalence of motivically flasque simplicial presheaves; it follows that $\mathcal{Z}$ is stably fibrant, and the level fibration is a stable fibration.

Since stable fibrations are closed under pullbacks, $\mathcal{Y} \times_{IQ_T J\mathcal{Y}} \mathcal{Z} \longrightarrow \mathcal{Z}$ is a stable fibration as well. Via part (ii) of Lemma 5.45 and $\mathcal{M}2$, the morphism $\mathcal{X} \longrightarrow \mathcal{Y} \times_{IQ_T J\mathcal{Y}} \mathcal{Z}$ is a stable equivalence. Factor this stable equivalence into a cofibration composed with a level fibration–level equivalence:

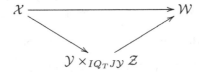

On account of 5.43, the cofibration in the diagram is a stable equivalence. On the other hand, the level fibration–level equivalence is a stable fibration according to 5.47. This proves the, in our case, non-trivial part of axiom $\mathcal{M}5$. □

The same definitions and arguments give the stable model structure for motivic spectra of spaces. We will leave the analogous formulations for spaces to the reader.

Finally, we shall interpret stable equivalences in terms of Nisnevich sheaves of bigraded stable homotopy groups of (s, t)-spectra. The first step is to make sense of the following statement:

Lemma 5.48 *There is a stable equivalence of motivic spectra*

$$E \longrightarrow E'$$

if and only if there is an isomorphism of motivic stable homotopy presheaves

$$\pi_{p,q}E \cong \pi_{p,q}E' .$$

When E is level fibrant, the starting point for defining motivic stable homotopy groups is the filtered colimit

$$\mathcal{E}_n \xrightarrow{\sigma_T^*} \Omega_T \mathcal{E}_{n+1} \xrightarrow{\Omega_T \sigma_T^*} \Omega_T^2 \mathcal{E}_{n+2} \xrightarrow{\Omega_T^2 \sigma_T^*} \cdots .$$

In X-sections, we get the groups $\pi_p Q_T \mathcal{E}_n(X)$ defined as the filtered colimit of the diagram

$$\pi_p(\mathcal{E}_n)(X) \xrightarrow{\pi_p(\sigma_T^*)(X)} \pi_p(\Omega_T \mathcal{E}_{n+1})(X) \xrightarrow{\pi_p(\Omega_T \sigma_T^*)(X)} \pi_p(\Omega_T^2 \mathcal{E}_{n+2})(X) \xrightarrow{\pi_p(\Omega_T^2 \sigma_T^*)(X)} \cdots .$$

Passing to the homotopy category associated to the motivic stable model structure over the scheme X, we can recast the latter as

$$[S_s^p, \mathcal{E}_n | X] \longrightarrow [S_s^p \wedge T, \mathcal{E}_{n+1} | X] \longrightarrow [S_s^p \wedge T^{\wedge 2}, \mathcal{E}_{n+2} | X] \longrightarrow \cdots .$$

Next, we want to rewrite this colimit taking into account the unstable version of 2.21.

Lemma 5.49 *There is a motivic weak equivalence between the Tate sphere T and the smash product $S_s^1 \wedge S_t^1$.*

Now, working in the homotopy category, so that we no longer need to impose the fibrancy condition, one obtains an alternative way of considering the groups in X-sections by taking the filtered colimit of the diagram

$$[S_s^p, \mathcal{E}_n | X] \longrightarrow [S_s^{p+1} \wedge S_t^1, \mathcal{E}_{n+1} | X] \longrightarrow [S_s^{p+2} \wedge S_t^2, \mathcal{E}_{n+2} | X] \longrightarrow \cdots .$$

Definition 5.50 *Let E be a motivic spectrum. The degree p and weight q motivic stable homotopy presheaf $\pi_{p,q}E$ is defined in X-sections by setting*

$$\pi_{p,q}E(X) := \operatorname*{colim}_{p,q \in \mathbb{Z}} \left([S_s^{p+n} \wedge S_t^{q+m}, \mathcal{E}_n | X] \longrightarrow [S_s^{p+n+1} \wedge S_t^{q+m+1}, \mathcal{E}_{n+1} | X] \cdots \right) .$$

Exercise 5.51 *Define*

$$\Omega_{S_s^1}(-) := \mathbf{Hom}_*(S_s^1, -) ,$$
$$\Omega_{S_t^1}(-) := \mathbf{Hom}_*(S_t^1, -) .$$

Show the presheaf isomorphisms

$$\pi_{p,q}E \cong \begin{cases} \pi_0 \Omega_{S^1_s}^{p-q} Q_T (JE[-q])_0 & p \geq q \,, \\ \pi_0 \Omega_{S^1_t}^{q-p} Q_T (JE[-p])_0 & p \leq q \,. \end{cases}$$

Unraveling the indices, one finds the identification

$$\pi_{p,q}E(X) \cong \pi_{p-q} Q_T J\mathcal{E}_{-q}(X) \,. \tag{27}$$

Next we look into the proof of 5.48.

Proof. A stable equivalence between E and E' induces for all integers $m \in \mathbb{Z}$ the levelwise schemewise weak equivalence

$$Q_T JE[m] \longrightarrow Q_T JE'[m] \,.$$

Hence, in all sections, the induced maps between motivic stable homotopy groups of E and E' are isomorphisms (27).

Conversely, if $\pi_{p,q}E$ and $\pi_{p,q}E'$ are isomorphic presheaves for $p \geq q \leq 0$, then there is a levelwise weak equivalence

$$Q_T JE \longrightarrow Q_T JE' \,.$$

This shows that $E \to E'$ is a stable equivalence.   □

Because of the motivic weak equivalence between T and $S^1_s \wedge S^1_t$, we may switch between motivic spectra and $S^1_s \wedge S^1_t$ spectra [Jar00, 2.13]. In other words, a motivic spectrum consists of pointed simplicial presheaves $\{\mathcal{E}_n\}_{n\geq 0}$ and structure maps

$$S^1_s \wedge S^1_t \wedge \mathcal{E}_n \longrightarrow \mathcal{E}_{n+1} \,.$$

Using this description, we shall see that a motivic spectrum E yields an (s,t)-bispectrum $E_{*,*}$ as discussed in the beginning of Sect. 2.3:

$$
\begin{array}{cccc}
\vdots & \vdots & \vdots & \\
S^2_t \wedge \mathcal{E}_0 & S^1_t \wedge \mathcal{E}_1 & \mathcal{E}_2 & \cdots \\
S^1_t \wedge \mathcal{E}_0 & \mathcal{E}_1 & S^1_s \wedge \mathcal{E}_2 & \cdots \\
\mathcal{E}_0 & S^1_s \wedge \mathcal{E}_1 & S^2_s \wedge \mathcal{E}_2 & \cdots
\end{array}
$$

In the s-direction there are structure maps

$$\sigma_s : S^1_s \wedge E_{m,n} \longrightarrow E_{m+1,n} \,.$$

If $m \geq n$, we use the identity morphism. If $m < n$, we use the morphism obtained from switching smash factors

$$S_s^1 \wedge S_t^n \wedge \mathcal{E}_m \xrightarrow{\tau \wedge 1} S_t^{n-1} \wedge S_s^1 \wedge S_t^1 \wedge \mathcal{E}_m \xrightarrow{1 \wedge \sigma} S_t^{n-1} \wedge \mathcal{E}_{m+1} .$$

Similarly, in the t-direction there are structure maps

$$\sigma_s : S_t^1 \wedge E_{m,n} \longrightarrow E_{m,n+1} .$$

If $m > n$, we use the morphism obtained from switching smash factors

$$S_t^1 \wedge S_s^m \wedge \mathcal{E}_n \xrightarrow{\tau \wedge 1} S_s^{m-1} \wedge S_t^1 \wedge S_s^1 \wedge \mathcal{E}_n \xrightarrow{1 \wedge \sigma} S_s^{m-1} \wedge \mathcal{E}_{n+1} .$$

If $m \leq n$, we use the identity morphism.

Associated to an (s,t)-bispectrum $E_{*,*}$, there are presheaves of bigraded stable homotopy groups $\pi_{p,q} E$. In X-sections, one considers the colimit of the diagram:

$$
\begin{array}{ccc}
\vdots & & \vdots \\
\Big\uparrow {\scriptstyle (\sigma_t)_*} & & \Big\uparrow {\scriptstyle (\sigma_t)_*} \\
[S_s^{p+m} \wedge S_t^{q+n+1}, E_{m,n+1}|X] \xrightarrow{(\sigma_s)_*} & [S_s^{p+m+1} \wedge S_t^{q+n+1}, E_{m+1,n+1}|X] & \longrightarrow \cdots \\
\Big\uparrow {\scriptstyle (\sigma_t)_*} & & \Big\uparrow {\scriptstyle (\sigma_t)_*} \\
[S_s^{p+m} \wedge S_t^{q+n}, E_{m,n}|X] \xrightarrow{(\sigma_s)_*} & [S_s^{p+m+1} \wedge S_t^{q+n}, E_{m+1,n}|X] & \longrightarrow \cdots
\end{array}
$$

Exercise 5.52 *In the above diagram, explain why there is no loss of generality in assuming that $E_{m,n}$ is motivically fibrant for all $m, n \in \mathbb{Z}$.*
Explicate the maps $(\sigma_s)_*$ and $(\sigma_t)_*$.

A cofinality argument shows the colimit of the above diagram of X-sections can be obtained by taking the diagonal and employing the transition maps $(\sigma_s)_*$ and $(\sigma_t)_*$ in either order. In particular, starting with a motivic spectrum, its degree p and weight q motivic stable homotopy presheaf is isomorphic to the bigraded presheaf $\pi_{p,q}$ of its associated (s,t)-bispectrum.

Lemma 5.48 and the previous observation show that

$$E \longrightarrow E'$$

is a stable equivalence if and only if there is an isomorphism of bigraded presheaves

$$\pi_{p,q} E_{*,*} \longrightarrow \pi_{p,q} E'_{*,*} .$$

The structure maps in the t-direction determine the sequence of morphisms of s-spectra

$$E_{*,0} \xrightarrow{(\sigma_t)_*} \Omega_{S_t^1} E_{*,1} \xrightarrow{\Omega_{S_t^1}(\sigma_t)_*} \Omega_{S_t^1}^2 E_{*,1} \xrightarrow{\Omega_{S_t^1}^2(\sigma_t)_*} \cdots .$$

The presheaf $\pi_{p,q}E$ is the filtered colimit of the presheaves in the diagram

$$s\pi_p \Omega_{S_t^1}^{q+n} JE_{*,n} \longrightarrow \pi_p \Omega_{S_t^1}^{q+n+1} JE_{*,n+1} \longrightarrow \cdots .$$

To conclude the discussion of homotopy groups, let E be a motivic spectrum, and note that a cofinality argument implies there is a natural isomorphism of bigraded presheaves

$$\pi_{p,q}E \cong \pi_{p,q}E_{*,*} .$$

References

[Bei87]   A. A. Beilinson. Height pairing between algebraic cycles. In *K-theory, arithmetic and geometry (Moscow, 1984–1986)*, vol. 1289 of *Lecture Notes in Math.*, pp. 1–25. Springer, Berlin, 1987.

[BF78]   A. K. Bousfield and E. M. Friedlander. Homotopy theory of Γ-spaces, spectra, and bisimplicial sets. In *Geometric applications of homotopy theory (Proc. Conf., Evanston, Ill., 1977), II*, vol. 658 of *Lecture Notes in Math.*, pp. 80–130. Springer, Berlin, 1978.

[BG73]   K. S. Brown and S. M. Gersten. Algebraic K-theory as generalized sheaf cohomology. In *Algebraic K-theory, I: Higher K-theories (Proc. Conf., Battelle Memorial Inst., Seattle, Wash., 1972)*, pp. 266–292. Lecture Notes in Math., Vol. 341. Springer, Berlin, 1973.

[BL95]   S. Bloch and S. Lichtenbaum. A spectral sequence for motivic cohomology. UIUC K-theory Preprint Archives, 62, 1995.

[Bla01]   B. A. Blander. Local projective model structures on simplicial presheaves. *K-Theory*, 24(3):283–301, 2001.

[Blo86]   S. Bloch. Algebraic cycles and higher K-theory. *Adv. Math.*, 61:267–304, 1986.

[DHI04]   D. Dugger, S. Hollander, and D. C. Isaksen. Hypercovers and simplicial presheaves. *Math. Proc. Cambridge Philos. Soc.*, 136(1):9–51, 2004.

[DI04]   D. Dugger and D. C. Isaksen. Weak equivalences of simplicial presheaves. In *Homotopy theory: relations with algebraic geometry, group cohomology, and algebraic K-theory*, vol. 346 of *Contemp. Math.*, pp. 97–113. Amer. Math. Soc., Providence, RI, 2004.

[DRØ03]   B. I. Dundas, O. Röndigs, and P. A. Østvær. Motivic functors. *Doc. Math.*, 8:489–525 (electronic), 2003.

[Dug01]   D. Dugger. Universal homotopy theories. *Adv. Math.*, 164(1):144–176, 2001.

[Dun]   B. I. Dundas. Nordfjordeid lectures. In *Summer school on motivic homotopy theory*. This volume.

[FS02]   E.M Friedlander and A. Suslin. The spectral sequence relating algebraic
           K-theory to motivic cohomology. *Ann. Sci. Ecole Norm. Sup.*, 35(6):773–
           875, 2002.

[GJ98]   P. G. Goerss and J. F. Jardine. Localization theories for simplicial
           presheaves. *Canad. J. Math.*, 50(5):1048–1089, 1998.

[Hir03]   P. S. Hirschhorn. *Model categories and their localizations*, vol. 99 of *Math-
           ematical Surveys and Monographs*. American Mathematical Society, Prov-
           idence, RI, 2003.

[Hov99]   M. Hovey. *Model categories*. American Mathematical Society, Providence,
           RI, 1999.

[Hov01]   M. Hovey. Spectra and symmetric spectra in general model categories. *J.
           Pure Appl. Algebra*, 165(1):63–127, 2001.

[HSS00]   M. Hovey, B. Shipley, and J. Smith. Symmetric spectra. *J. Amer. Math.
           Soc.*, 13(1):149–208, 2000.

[Isa04]   D. Isaksen. Flasque model structures for presheaves. UIUC K-theory
           Preprint Archives, 679, 2004.

[Jar87]   J. F. Jardine. Simplicial presheaves. *J. Pure Appl. Algebra*, 47(1):35–87,
           1987.

[Jar00]   J. F. Jardine. Motivic symmetric spectra. *Doc. Math.*, 5:445–553 (elec-
           tronic), 2000.

[Jar03]   J. F. Jardine. Intermediate model structures for simplicial presheaves.
           Preprint, 2003.

[Lár04]   F. Lárusson. Model structures and the Oka principle. *J. Pure Appl.
           Algebra*, 192(1–3):203–223, 2004.

[Lev]   M. Levine. Nordfjordeid lectures. In *Summer school on motivic homotopy
           theory*. This volume.

[Lev01]   M. Levine. Techniques of localization in the theory of algebraic cycles. *J.
           Algebraic Geom.*, 10(2):299–363, 2001.

[Lev03]   M. Levine. The homotopy coniveau filtration. UIUC K-theory Preprint
           Archives, 628, 2003.

[MV99]   F. Morel and V. Voevodsky. $\mathbb{A}^1$-homotopy theory of schemes. *Inst. Hautes
           Études Sci. Publ. Math.*, (90):45–143 (2001), 1999.

[Nee96]   A. Neeman. The Grothendieck duality theorem via Bousfield's techniques
           and Brown representability. *J. Amer. Math. Soc.*, 9(1):205–236, 1996.

[Nee01]   A. Neeman. *Triangulated categories*, vol. 148 of *Annals of Mathematics
           Studies*. Princeton University Press, Princeton, NJ, 2001.

[Smi]   J. Smith. Combinatorial model categories. In preparation.

[Sus03]   A. Suslin. On the Grayson spectral sequence. In *Number theory, algebra,
           and algebraic geometry. Collected papers dedicated to the 80th birthday of
           Academician Igor' Rostislavovich Shafarevich. Transl. from the Russian.
           Moskva: Maik Nauka/Interperiodika. Proceedings of the Steklov Institute
           of Mathematics* 241, 202–237 (2003); translation from *Tr. Mat. Inst. Im.
           V. A. Steklova* 241, 218–253, 2003.

[SV96]   A. Suslin and V. Voevodsky. Singular homology of abstract algebraic
           varieties. *Invent. Math.*, 123(1):61–94, 1996.

[Voe98]   V. Voevodsky. $\mathbb{A}^1$-homotopy theory. In *Proceedings of the International
           Congress of Mathematicians, Vol. I (Berlin, 1998)*, pp. 579–604 (elec-
           tronic), 1998.

[Voe00a] V. Voevodsky. Homotopy theory of simplicial sheaves in completely de-composable topologies. UIUC K-theory Preprint Archives, 443, August 2000.

[Voe00b] V. Voevodsky. Triangulated categories of motives over a field. In *Cycles, transfers, and motivic homology theories*, pp. 188–238. Princeton Univ. Press, Princeton, NJ, 2000.

[Voe02a] V. Voevodsky. Cancellation theorem. UIUC K-theory Preprint Archives, 541, 2002.

[Voe02b] V. Voevodsky. Open problems in the motivic stable homotopy theory. I. In *Bogomolov, Fedor (ed.) et al., Motives, polylogarithms and Hodge theory. Part I: Motives and polylogarithms. Papers from the International Press conference, Irvine, CA, USA, June 1998. Somerville, MA: International Press. Int. Press Lect. Ser.* 3, No. I, 3–34. 2002.

[Voe02c] V. Voevodsky. A possible new approach to the motivic spectral sequence for algebraic K-theory. In *Recent progress in homotopy theory (Baltimore, MD, 2000)*, vol. 293 of *Contemp. Math.*, pp. 371–379. Amer. Math. Soc., Providence, RI, 2002.

Index

Universitext

Aguilar, M.; Gitler, S.; Prieto, C.: Algebraic Topology from a Homotopical Viewpoint

Aksoy, A.; Khamsi, M. A.: Methods in Fixed Point Theory

Alevras, D.; Padberg M. W.: Linear Optimization and Extensions

Andersson, M.: Topics in Complex Analysis

Aoki, M.: State Space Modeling of Time Series

Arnold, V. I.: Lectures on Partial Differential Equations

Audin, M.: Geometry

Aupetit, B.: A Primer on Spectral Theory

Bachem, A.; Kern, W.: Linear Programming Duality

Bachmann, G.; Narici, L.; Beckenstein, E.: Fourier and Wavelet Analysis

Badescu, L.: Algebraic Surfaces

Balakrishnan, R.; Ranganathan, K.: A Textbook of Graph Theory

Balser, W.: Formal Power Series and Linear Systems of Meromorphic Ordinary Differential Equations

Bapat, R.B.: Linear Algebra and Linear Models

Benedetti, R.; Petronio, C.: Lectures on Hyperbolic Geometry

Benth, F. E.: Option Theory with Stochastic Analysis

Berberian, S. K.: Fundamentals of Real Analysis

Berger, M.: Geometry I, and II

Bliedtner, J.; Hansen, W.: Potential Theory

Blowey, J. F.; Coleman, J. P.; Craig, A. W. (Eds.): Theory and Numerics of Differential Equations

Blowey, J.; Craig, A.: Frontiers in Numerical Analysis. Durham 2004

Blyth, T. S.: Lattices and Ordered Algebraic Structures

Börger, E.; Grädel, E.; Gurevich, Y.: The Classical Decision Problem

Böttcher, A; Silbermann, B.: Introduction to Large Truncated Toeplitz Matrices

Boltyanski, V.; Martini, H.; Soltan, P. S.: Excursions into Combinatorial Geometry

Boltyanskii, V. G.; Efremovich, V. A.: Intuitive Combinatorial Topology

Bonnans, J. F.; Gilbert, J. C.; Lemaréchal, C.; Sagastizábal, C. A.: Numerical Optimization

Booss, B.; Bleecker, D. D.: Topology and Analysis

Borkar, V. S.: Probability Theory

Brunt B. van: The Calculus of Variations

Bühlmann, H.; Gisler, A.: A Course in Credibility Theory and its Applications

Carleson, L.; Gamelin, T. W.: Complex Dynamics

Cecil, T. E.: Lie Sphere Geometry: With Applications of Submanifolds

Chae, S. B.: Lebesgue Integration

Chandrasekharan, K.: Classical Fourier Transform

Charlap, L. S.: Bieberbach Groups and Flat Manifolds

Chern, S.: Complex Manifolds without Potential Theory

Chorin, A. J.; Marsden, J. E.: Mathematical Introduction to Fluid Mechanics

Cohn, H.: A Classical Invitation to Algebraic Numbers and Class Fields

Curtis, M. L.: Abstract Linear Algebra

Curtis, M. L.: Matrix Groups

Cyganowski, S.; Kloeden, P.; Ombach, J.: From Elementary Probability to Stochastic Differential Equations with MAPLE

Dalen, D. van: Logic and Structure

Das, A.: The Special Theory of Relativity: A Mathematical Exposition

Debarre, O.: Higher-Dimensional Algebraic Geometry

Deitmar, A.: A First Course in Harmonic Analysis

Demazure, M.: Bifurcations and Catastrophes

Devlin, K. J.: Fundamentals of Contemporary Set Theory

DiBenedetto, E.: Degenerate Parabolic Equations

Diener, F.; Diener, M.(Eds.): Nonstandard Analysis in Practice

Dimca, A.: Sheaves in Topology

Dimca, A.: Singularities and Topology of Hypersurfaces

DoCarmo, M. P.: Differential Forms and Applications

Duistermaat, J. J.; Kolk, J. A. C.: Lie Groups

Dundas, B. I., Levine, M., Voevodsky, V., Østvær, P. A., Röndigs, O., Jahren, B.: Motivic Homotopy Theory

Edwards, R. E.: A Formal Background to Higher Mathematics Ia, and Ib

Edwards, R. E.: A Formal Background to Higher Mathematics IIa, and IIb

Emery, M.: Stochastic Calculus in Manifolds

Emmanouil, I.: Idempotent Matrices over Complex Group Algebras

Endler, O.: Valuation Theory

Erez, B.: Galois Modules in Arithmetic

Everest, G.; Ward, T.: Heights of Polynomials and Entropy in Algebraic Dynamics

Farenick, D. R.: Algebras of Linear Transformations

Foulds, L. R.: Graph Theory Applications

Franke, J.; Härdle, W.; Hafner, C. M.: Statistics of Financial Markets: An Introduction

Frauenthal, J. C.: Mathematical Modeling in Epidemiology

Freitag, E.; Busam, R.: Complex Analysis

Friedman, R.: Algebraic Surfaces and Holomorphic Vector Bundles

Fuks, D. B.; Rokhlin, V. A.: Beginner's Course in Topology

Fuhrmann, P. A.: A Polynomial Approach to Linear Algebra

Gallot, S.; Hulin, D.; Lafontaine, J.: Riemannian Geometry

Gardiner, C. F.: A First Course in Group Theory

Gårding, L.; Tambour, T.: Algebra for Computer Science

Gärtner B.; Matoušek J.: Understanding and Using Linear Programming

Godbillon, C.: Dynamical Systems on Surfaces

Godement, R.: Analysis I, and II

Godement, R.: Analysis II

Goldblatt, R.: Orthogonality and Spacetime Geometry

Gouvêa, F. Q.: p-Adic Numbers

Gross, M. et al.: Calabi-Yau Manifolds and Related Geometries

Gustafson, K. E.; Rao, D. K. M.: Numerical Range. The Field of Values of Linear Operators and Matrices

Gustafson, S. J.; Sigal, I. M.: Mathematical Concepts of Quantum Mechanics

Hahn, A. J.: Quadratic Algebras, Clifford Algebras, and Arithmetic Witt Groups

Hájek, P.; Havránek, T.: Mechanizing Hypothesis Formation

Heinonen, J.: Lectures on Analysis on Metric Spaces

Hlawka, E.; Schoißengeier, J.; Taschner, R.: Geometric and Analytic Number Theory

Holmgren, R. A.: A First Course in Discrete Dynamical Systems

Howe, R., Tan, E. Ch.: Non-Abelian Harmonic Analysis

Howes, N. R.: Modern Analysis and Topology

Hsieh, P.-F.; Sibuya, Y. (Eds.): Basic Theory of Ordinary Differential Equations

Humi, M., Miller, W.: Second Course in Ordinary Differential Equations for Scientists and Engineers

Hurwitz, A.; Kritikos, N.: Lectures on Number Theory

Huybrechts, D.: Complex Geometry: An Introduction

Isaev, A.: Introduction to Mathematical Methods in Bioinformatics

Istas, J.: Mathematical Modeling for the Life Sciences

Iversen, B.: Cohomology of Sheaves

Jacod, J.; Protter, P.: Probability Essentials

Jennings, G. A.: Modern Geometry with Applications

Jones, A.; Morris, S. A.; Pearson, K. R.: Abstract Algebra and Famous Inpossibilities

Jost, J.: Compact Riemann Surfaces

Jost, J.: Dynamical Systems. Examples of Complex Behaviour

Jost, J.: Postmodern Analysis

Jost, J.: Riemannian Geometry and Geometric Analysis

Kac, V.; Cheung, P.: Quantum Calculus

Kannan, R.; Krueger, C. K.: Advanced Analysis on the Real Line

Kelly, P.; Matthews, G.: The Non-Euclidean Hyperbolic Plane

Kempf, G.: Complex Abelian Varieties and Theta Functions